Pitman Research Notes in Mathematics Series

Submission of proposals for consideration

Suggestions for publication, in the form of outlines and representative samples, are invited by the Editorial Board for assessment. Intending authors should approach one of the main editors or another member of the Editorial Board, citing the relevant AMS subject classifications. Alternatively, outlines may be sent directly to the publisher's offices. Refereeing is by members of the board and other mathematical authorities in the topic concerned, throughout the world.

Preparation of accepted manuscripts

On acceptance of a proposal, the publisher will supply full instructions for the preparation of manuscripts in a form suitable for direct photo-lithographic reproduction. Specially printed grid sheets can be provided and a contribution is offered by the publisher towards the cost of typing. Word processor output, subject to the publisher's approval, is also acceptable.

Illustrations should be prepared by the authors, ready for direct reproduction without further improvement. The use of hand-drawn symbols should be avoided wherever possible, in order to maintain maximum clarity of the text.

The publisher will be pleased to give any guidance necessary during the preparation of a typescript, and will be happy to answer any queries.

Important note

In order to avoid later retyping, intending authors are strongly urged not to begin final preparation of a typescript before receiving the publisher's guidelines. In this way it is hoped to preserve the uniform appearance of the series.

Addison Wesley Longman Ltd
Edinburgh Gate
Harlow, Essex, CM20 2JE
UK
(Telephone (0) 1279 623623)

Titles in this series. A full list is available from the publisher on request.

John W Schutz

La Trobe University, Australia

Independent axioms for Minkowski space-time

 LONGMAN

Addison Wesley Longman Limited
Edinburgh Gate, Harlow
Essex CM20 2JE, England
and Associated Companies throughout the world.

*Published in the United States of America
by Addison Wesley Longman Inc.*

First published 1997

AMS Subject Classifications   (Main) 51-02, 83A05, 51P05
                               (Subsidiary) 51M05, 51A99, 03H10

ISSN 0269-3674

ISBN 0 582 31760 6

British Library Cataloguing in Publication Data

A catalogue record for this book is
available from the British Library

Printed and bound in Great Britain
by Biddles Ltd, Guildford and King's Lynn

To my wife

my parents

and my children

Amina

Edith and Steven

Helena and Clara

Contents

Appendices

Preface

The primary aim of this monograph is to clarify the undefined primitive concepts and the axioms which form the basis of Einstein's theory of special relativity. Minkowski space–time is developed from a set of independent axioms stated in terms of a single relation of intermediacy or betweenness. It is shown that all models are isomorphic to the usual coordinate model and the axioms are consistent relative to the reals.

The investigation has revealed some open questions which are described in an appendix.

In a lecture delivered before the International Congress of Mathematicians at Paris in 1900, Hilbert (1900) suggested:

> " .   .   .   that wherever, from the side of the theory of knowledge or in geometry, or from the theories of natural or physical science, mathematical ideas come up, the problem arises for mathematical science to investigate the principles underlying these ideas and so to establish them upon a simple and complete system of axioms, . . . "

and also stated several mathematical problems, the sixth problem being

> "Investigations of the foundations of geometry suggest the problem: to treat in the same manner, by means of axioms, those physical sciences in which mathematics plays an important part; first of all, the theory of probability and mechanics."

The present investigation of Minkowski space–time is more closely related to investigations into geometry than to physics. However the additional assumptions required for relativistic mechanics are few and the development of the theory is straightforward once the nature and properties of the space–time have been described.

Penrose (1989) has formulated three broad categories of physical theories: "superb", "useful" and "tentative" where a theory is categorised as "superb" if it applies over a wide range of situations with high accuracy. These criteria are clearly satisfied by Euclidean geometry and Minkowski space–time which are generally regarded as geometric theories but which may also be regarded as "superb" theories of physics since one is subsumed by Newtonian mechanics and the other by Einstein's theory of special relativity. The category of "superb" theories of physics includes Euclidean geometry, Newtonian mechanics, Maxwell's theory of electromagnetism, Einstein's theories of special and general relativity, which subsume Minkowski space–time and Riemannian geometry, as well as quantum theory and also, possibly, quantum electrodynamics. Thus Minkowski space–time may be considered as both a geometric theory and also as a theory of physics.

One might expect that Minkowski space–time would require a system of axioms and primitive concepts even more complicated than those for Euclidean geometry since the space–time is a more complicated structure than the geometry. Yet, surprisingly, in some ways the axiomatic system presented here for Minkowski space–time appears to be simpler than the corresponding axiomatic systems for Euclidean geometry. Only one undefined relation is required for Minkowski space–time whereas two are required for Euclidean geometry. It is the additional structure, namely "light signals", which permits the concept of congruence to be derived within the theory for Minkowski space–time, whereas for Euclidean geometry the concept of congruence must be stated as an additional primitive undefined relation. Comparisons between the systems of axioms are quite favourable but not as straightforward.

A frequently–asked question, related to the question of comparative simplicity of the systems of axioms, is: How many independent axioms are required for Euclidean geometry and Minkowski space–time? There is no definitive answer to this question as differences of style affect the choice of postulated properties and the manner of their expression. While it is commonly believed that Euclidean geometry requires two primitive relations and five postulates, there are many assumed properties in the original exposition of Euclid. Hilbert (1899) expressed Euclidean geometry in terms of two undefined sets of *points* and *lines*, the same two primitive relations and five independent groupings of axioms containing a total of twenty non–independent axioms. The system of Hilbert was subsequently developed further by Veblen (1904) and then by Moore (1908) whose system of independent axioms was expressed in terms of a single undefined set of *points*, the same two undefined relations and fifteen independent axioms. The present axiomatic system for Minkowski space–time is expressed in terms of two undefined sets of *events* and *paths*, a single three–point relation of *intermediacy* or *betweenness* and fifteen independent axioms. Thus, in comparison with the axiomatic systems of Veblen (1904) and Moore (1908), it could be claimed that the present axiomatic system for Minkowski space–time is no more complicated than the axiomatic systems for Euclidean geometry. Indeed the primitive basis for Minkowski space–time has but a single relation which is certainly simpler than the basis of two relations for Euclidean geometry. Issues related to questions of "comparative simplicity" are discussed in Chapters 1 and 2 and Appendix 4.

In the present axiomatic system the statements of several axioms are somewhat complicated because of the formal requirement to be able to demonstrate independence models or counterexamples. To clarify the meaning and significance of the axioms, the formal requirement of independence can be relaxed to provide a corresponding list of non–independent axioms which are stated in Appendix 2. In this alternative axiomatic system the order properties and kinematic relations are clearer and more intuitively acceptable.

The present axiomatic system is a significant development from the previous axiomatic systems of Schutz (1973, 1981) since each of the present axioms is justified by a proof of its independence. It is in this sense that it could be claimed that the assumptions of the present system are minimal. Unlike its predecessors (Walker (1948, 1949), Szekeres (1968), Schutz (1973, 1981)), the present axiomatic system makes no assumptions related to "the direction of time", "causality" or the existence of "light signals". The undefined concepts and axioms of the theory are therefore simpler, although a considerable theoretical development is required to obtain results which were assumed in the previous axiomatic systems. Many new intermediate results are required and many previous results are now obtained in a different sequence and often with substantially different proofs. Much of the development of Chapters 3 to 5 is devoted to establishing the existence of "light signals" and properties related to a "direction of time". This is well illustrated by the Causality Theorem 26 which is a consequence of twenty five preceding theorems, whereas in the previous axiomatic systems the analogous proposition was an axiom. The present weaker system of axioms has necessitated a completely different approach for the extension of results to the full $3 + 1$–dimensional space–time. The only chapter with minor modifications is Chapter 8 on one–dimensional kinematics.

Many of the theorems and proofs of the present treatment are based on earlier results of Veblen (1904, 1911) for affine geometry and Walker (1948, 1959) for cosmology. Even though the present axiomatic system bears little resemblance to that of Walker, many of the results and concepts are based on the excellent analyses of the concepts of "collinearity" and "optical lines" given by Walker and are acknowledged in the text.

I wish to acknowledge assistance and encouragement from many people. George Szekeres offerred many constructive comments, emphasized the relevance of the properties of parallelism for one–dimensional kinematics and encouraged me to develop the theory of Chapter 7. It is a pleasure to thank many people for encouragement and stimulating discussions including Reinhard Breuer, Jürgen Ehlers, Ted Fackerell, Lloyd Humberstone, Richard Josza, Erwin Kronheimer, Roger Penrose, Michael Streubel, Patrick Suppes, Esther Szekeres, Paul Tod, Florence Tsou and Nicholas Woodhouse. It is also a pleasure to thank Graeme Byrne, Phillip Rice and Philip Scott for assistance with the software and Michele Cartwright, Claire Hollingsworth, Heather Koch and Stephen West for assistance with the typesetting.

1. Introduction

1.1 Axiomatic systems

The main purpose of describing a theory axiomatically is to clarify the basic concepts of the theory and the assumptions on which it is based.

Axiomatic theories for geometries and space–times usually presuppose some other theories. For the present system of axioms for Minkowski space–time, we will presuppose mathematical logic, set theory and the theory of the reals. We will show that:

- all models of the axiomatic system are isomorphic (Section 9.5),
- the axiomatic system is consistent relative to the reals (Section 10.6), and
- the axioms are mutually independent (Chapter 11).

The system of axioms is stated in terms of a single undefined three–term relation of "betweenness" or "intermediacy".

1.2 Independence and consistency of the set of axioms

In Section 10.6 we will show that the set of axioms is consistent relative to the reals, by showing that all the axioms are satisfied in the usual model of Minkowski space–time.

A desirable property of an axiomatic system is that the assumptions are minimal in the sense of the properties or structure which they assume. One way to establish this is show that the axioms are mutually independent.

For each axiom, we demonstrate its independence by describing a model which satisfies all other axioms but not the given axiom. Thus the existence of the model establishes that the statement of the given axiom can *not* be proved as a theorem from the remaining axioms. Another way of stating this is to say that the existence of the independence model provides us with the justification to regard the given proposition as an axiom. Each independence model is a counterexample which serves to illustrate why the statement of the axiom is required for the categorical[1] description of Minkowski space–time.

---

[1] Notes to the chapters appear in Appendix 4 (p.225).

A classic example of the application of this method is provided by the Cayley–Klein model of Bolyai–Lobachevski geometry, which demonstrates the independence of the Euclidean parallel postulate from the other axioms of Euclidean geometry.

The detailed descriptions of the independence models are given in Chapter 11. For Axiom O4 and Axiom C there are, respectively, two and three independence models: their properties emphasize the significance of different aspects of the statements of the axioms.

1.3 Axiomatic systems for geometries

Many axiomatic systems have been given for geometries and several are reviewed by Torretti (1978) and Suppes *et al.* (1989). The earliest approach due to Euclid has been further developed by Pasch (1882), Peano (1889), Hilbert (1899) and others.

A second tradition is to characterize geometries by properties of symmetry and free mobility: this is known as the Helmholtz–Lie space form problem and its solutions are discussed and reviewed by Busemann (1955) and Freudenthal (1965). Most of the solutions are stated within a context of differentiable manifolds or spaces satisfying various topological and geometric axioms — an exception is the system of axioms using symmetry given by Rédei (1968) which is closer to the tradition of Euclid and Hilbert.

Most of the axiomatic systems for these geometries are within these two traditions. Some other axiomatic systems are reviewed by Torretti (1978).

Independent axiomatic systems Hilbert (1899) demonstrated the independence of the five major groupings of axioms of Euclidean geometry so as to illustrate the significance of the conclusions that can be drawn from the individual axioms.

Veblen (1904) has presented a system of independent axioms for affine geometry and has shown that many non–isomorphic Euclidean metrics may be imposed by reference to polarities in the plane at infinity. Moore (1908) has extended and modified the earlier system of Veblen (1904) and has shown that Euclidean geometry may be developed from fifteen independent axioms. Veblen and Young (1908) have shown that several projective geometries, in particular the real and complex projective geometries, have categorical systems of independent axioms.

For any given mathematical structure, the number of axioms is related to the selection of undefined terms and relations. Issues affecting the choice of a primitive or undefined basis are discussed further in the notes to Chapter 2 and the open questions of Appendix 3.

1.4 Axiomatic systems for space–times

Minkowski space–time has a more complicated structure than Euclidean geometry and this may explain the diversity of axiomatic approaches, which is in marked contrast to the two traditions for the Euclidean and non–Euclidean geometries. Minkowski space–time has three distinct types of interval — timelike, spacelike and lightlike — and the lightlike intervals enable comparisons to be made between other types of interval, so that "light signals" can be used to define a relation of congruence. This was done in one of the earliest axiomatic systems by Robb (1936) whose axiomatic system was stated in terms of a single undefined relation of temporal order. However Robb's aim of "mathematical simplicity" was achieved with a minimal number of primitive concepts rather than by the intuitive acceptability, minimal number, or independence of the axioms. Robb's approach has been further developed by Mundy (1986a).

The Helmholtz–Lie tradition   The space–time problem analogous to the Helmholtz–Lie space form problem has been investigated by Busemann (1967), Freudenthal (1964) and Mayr (1983) in wider contexts of space–time structures which are assumed to satisfy various topological axioms . Busemann (1967) considers a property of "symmetry of the light cone", which also appears in the axiomatic system of Guts and Levichev (1984) together with an axiom of space–time "homogeneity". Freudenthal (1964) and Mayr (1983) use a property of "free mobility".

Affine space–times   Some authors assume the concept of an affine space. Aleksandrov (1969) considers a symmetry property which he describes as "isotropy of the light cone", together with additional conditions, to show that the light cones are ellipsoidal cones in an affine space, while Alexandrov (1967) uses a property of "reflection in the path of every observer". Pimenov (1968) discusses "affine space–times" with a property of "mobility of frames". Other axiomatic systems have been discussed by Mundy (1986a,b).

First–order axiomatic systems   Goldblatt (1987) has presented a first–order axiomatic system for Minkowski space–time in terms of primitive relations of betweennness and orthogonality. Goldblatt's aim is to develop a theory which is deductively complete and decidable, however the theory is not categorical since it has non–isomorphic models. The system of Goldblatt could, perhaps, be further refined by considering whether some of the axioms might be obtained as theorems.

The coordinate frame approach   Other authors assume the concept of a coordinate frame; in particular, Bunge (1967) has axiomatized the conventional approach due to Einstein (1905), while Suppes (1954, 1959) and Noll (1964) have based their systems on the assumption of the invariance of the quadratic form

$$\Delta x_1^2 + \Delta x_2^2 + \Delta x_3^2 - c^2 \Delta t^2$$

with respect to transformations between coordinate frames.

The present system and its predecessors   Axiomatic systems using properties of symmetry together with properties of the paths of freely-moving observers, have been proposed by several authors. Walker (1948, 1959) considered the foundations for an approach to relativistic cosmology which had been developed by Milne and Whitrow (Milne, 1948) to describe space–times in terms of observations which could be made using direct and reflected light signals together with local time measurements. The undefined basis involved the concepts of "particles", "light signals", and "temporal order". Walker's axiom system was not developed sufficiently to describe Minkowski space–time and was, in fact, restricted to sets of relatively stationary particles, but it succeeded in clarifying many kinematic concepts, especially those of "particle" (or "path" in the present system), "light signal", "optical line" and "collinearity".

Walker (1959) discusses cosmologies of "fundamental particles" which satisfy two axioms of symmetry, one of which is a property of isotropy similar to that of the present treatment. The axiomatic system of Szekeres (1968) has an axiom of isotropy and two other symmetry axioms. Schutz (1973, 1981) uses a property of isotropy similar to that of the present treatment, but characterizes Minkowski space–time indirectly by first demonstrating that its velocity space is hyperbolic. Of these axiomatic systems, only those of Schutz (1973) and Szekeres (1968) describe Minkowski space–time in terms of assumptions related to what one might describe as either "kinematic experience" or physical intuition.

The undefined basis of the present system bears some resemblance in an intuitive or physical sense to the bases used by Walker (1948, 1949), Szekeres (1968) and Schutz (1973). Walker (1948, 1959) considers "particles" as subsets of events with two undefined relations of order, one being a "signal relation" and the other a relation of "temporal order" for each particle. The undefined basis of Szekeres (1968) bears some resemblance to that of Walker, although Szekeres regards both particles and light signals as linearly ordered subsets which are assumed to be order–isomorphic to the reals. The previous axiomatic system of the author (Schutz, 1973) has an undefined basis which is similar to that of Szekeres (1968) in a physical sense and to that of Walker (1948, 1959) in a formal sense. "Particles" were defined as subsets of the set of events with a primitive or undefined "signal relation" on the set of events which then imposed a partial order called a "temporal order" relation.

Many of the theorems and proofs of the present treatment are based on earlier results of Veblen (1904, 1911) for affine geometry, Walker (1948, 1949) for cosmology, and Schutz (1973, 1981) for Minkowski space–time.

Useful surveys of the literature on characterizations of Minkowski space–time and the Lorentz group are given by Guts (1982) and Suppes *et al.* (1989).

Axiomatic systems for curved space–times   More general space–times, such as those of general relativity and other gravitational and unified field theories, have a considerable amount of structure which involves causal, topological, differentiable, conformal, projective and pseudo–metric properties. Early attempts at axiomatizing general relativity were proposed by Reichenbach (1924) and Weyl (1918).

Kronheimer and Penrose (1967) discussed general cases of space–times which satisfy causal properties. This approach has been further developed by Kronheimer (1971) and Carter (1971). Busemann (1967), Pimenov (1968, 1988) and Szekeres (1991, 1994) have considered space–times which satisfy causal and various topological axioms.

Castagnino (1971) axiomatized space–time by developing the metric structure from consideration of the paths of particles and light signals with methods previously developed by Kundt and Hoffmann (1962) and Marzke and Wheeler (1964).

Ehlers, Pirani and Schild (1972) have given a detailed axiomatic discussion of the topological, differentiable, conformal, projective, affine and metric structures. This has been further developed by Ehlers (1973), Ehlers and Köhler (1977), Ehlers and Schild (1973), Pirani (1973), Woodhouse (1973), Coleman and Korte (1984) and Castagnino and Ordóñez (1989).

1.5 A brief introduction to the present axiomatic system

The present axiomatic system may be considered to follow from the two traditional approaches to the absolute geometries. The first tradition of Euclid–Hilbert–Veblen–Moore can be clearly seen in most of the axioms, apart from the Axiom of Symmetry (Axiom S) which expresses a property of mobility and is more in the spirit of the Helmholtz–Lie tradition. The two traditions of geometry are combined for Minkowski space–time which is expressed in terms of two undefined sets of *events* and *paths*, a single three–point *betweenness* relation and fifteen independent axioms. The axiomatic system could have been stated in terms of a single undefined set of *events*, the same *betweenness* relation and thirteen independent axioms. While this alternative may appear to be "simpler" and therefore attractive, there are several good reasons for the choice of a system of fifteen axioms and some of these reasons are discussed in the notes 1–4 to Chapter 2 (p.225).

Many of the axioms of the present system[2] bear a strong resemblance to those of Hilbert (1899), Veblen (1904, 1911) and other writers. The exceptions are Axioms I5—I7 (of incidence) which distinguish a space–time from a geometry and Axiom S (of symmetry) which replaces the congruence postulates for the geometries. Apart from these exceptions the axioms of the present system have counterparts in the axiom systems for the geometries. All these axiomatic systems are second–order theories.

The present axiomatic system is second–order since it includes a second–order statement of the Axiom of Continuity. There are several reasons for choosing a second–order axiom system: the main reason is explained in Appendix 3 where it is observed that replacement of the second–order Axiom of Continuity by an infinite schema of first–order axioms leads to some space–time models which bear a closer "physical resemblance" to Galilean space–time than to Minkowski space–time. There are many unanswered open questions remaining for first–order axiomatizations of Minkowski space–time and some are described in Appendix 3.

There are several assumptions which we do not make. In particular, as in the previous axiomatic system of Schutz (1973, 1981), we do not assume the concept of a coordinate frame, we do not assume that the events of a path can be ordered by the real numbers, nor do we assume that paths and light signals move with constant speed. As a further development from the previous axiomatic systems, we do not assume the existence of a temporal order relation on the set of events.

Three properties of Minkowski space–time are of central importance to the subsequent development. Firstly, one–dimensional kinematics is in many ways analogous to plane absolute geometry except for one significant difference, namely the existence of "unreachable sets". Secondly, the most important analogous property is parallelism and the corresponding question of uniqueness of parallelism is closely related to uniform motion along paths. Both Robb (1936) and Szekeres (1968) observed that uniform motion implies uniqueness of parallelism but, as in the previous system of Schutz (1973), we are able to prove the uniqueness of parallelism and then show that this implies uniform motion along paths, so that we do not need to assume Newton's first law of motion explicitly. Thirdly, in contrast with the Euclidean velocity space of Newtonian kinematics, the velocity space associated with Minkowski space–time is hyperbolic, a property which is established in the present treatment using a characterization of ellipsoids in affine spaces due to Busemann (1955). This characterization is much more readily acccessible than the results used in the previous axiomatic systems of Schutz (1973, 1981).

In the previous axiomatic systems of Schutz (1973, 1981) the primary aim was to clarify the foundations of Special Relativity so that the theory would become as acceptable and familiar as Euclidean geometry. The present axiomatic system has resulted from the more stringent requirement of independence of the axioms. The search for independent axioms has imposed a need for further clarification of the undefined basis of primitive notions (which now consists of undefined elements called "events", certain subsets of the set of events called "paths" and a single undefined relation of "betweenness"), and has resulted in axioms which are more intuitively acceptable.

We are able to demonstrate that the system of axioms is categorical by demonstrating in Chapter 9 that all models are isomorphic to the usual cooordinatized model of Minkowski space–time. The existence of this model also serves to establish that the system of axioms is consistent or, to be more precise, consistent relative to the reals.

The requirement of independence of the axioms provides exacting criteria to justify the name of "axiom" for each corresponding assumption. Independence of an axiom is proved by the existence of the corresponding independence model or counterexample, which demonstrates the impossibility of proving the statement of the axiom as a theorem. However the statements of some of the axioms are somewhat complicated because of the formal requirement to be able to demonstrate independence models or counterexamples. This tends to obscure the intuitive meaning and significance of some of the axioms. To clarify the meaning of the axioms, the formal requirement of independence can be relaxed to provide a corresponding list of non–independent axioms which are stated as the Alternative Axiom System of Appendix 2. In this Alternative Axiom System the order properties and kinematic relations become much clearer and more intuitively acceptable.

Since much of the terminology and notation is new, a listing of the definitions and notation is included after the appendices.

2. Primitive notions and axioms

In Section 2.1 we specify the primitive undefined basis. The system of axioms consists of six axioms of order (Section 2.2), seven axioms of incidence (Section 2.3), one axiom of symmetry or isotropy (Section 2.4), and one axiom of continuity (Section 2.5). The axioms of order, some of the axioms of incidence and the axiom of continuity resemble axioms used for Euclidean geometry by Hilbert (1899), Veblen (1904, 1911) and Moore (1908), as well as axioms for real projective geometry given by Veblen and Young (1908, 1918).

For Euclidean geometry the metric properties are a consequence, in the systems of Hilbert (1899), Moore (1908) and Veblen (1911), of further axioms of congruence and an axiom of parallels, while in Veblen's (1904) system they depend upon an axiom of parallels and the specification of a polarity. Minkowski space–time has a richer structure than Euclidean geometry since there are some pairs of events which can be joined by single paths and other pairs of events which can not be joined by single paths. This additional structure is specified by the fifth, sixth and seventh axioms of incidence which could, perhaps, be called "axioms of ordered non-incidence". As a consequence of its richer structure Minkowski space–time can be specified categorically without an axiom of uniqueness of parallelism, without axioms of congruence and without reference to a polarity.

2.1 Primitive undefined basis[1]

Minkowski space time is

$$\mathcal{M} = \langle \mathcal{E}, \mathcal{P}, [\cdots] \rangle$$

where $\mathcal{E}$ is a set whose elements are called *events*, $\mathcal{P}$ is a set of subsets of $\mathcal{E}$ called *paths* and $[\cdots]$ is a ternary relation on the set of events of $\mathcal{E}$ called a *betweenness relation*.

Paths will be denoted by upper case symbols $Q, R, S, \cdots$; events will be denoted by lower case symbols $a, b, c, d, \cdots$ or by a path symbol with a subscript, such as for example Q_a, Q_1, Q_x, Q_α for events which belong to the path Q and $R_b, R_3, R_w R_\gamma$ for events which belong to the path R. Given a pair of distinct events $a, b \in S$ we say that a, b *belong to S* or *lie on S* or that they can *be connected by S* or that S *passes* *through* them.

We will presuppose logic, set theory and the arithmetic of real numbers.

2.2 Axioms of order

The axioms of order resemble axioms or theorems of the geometric axiom systems of Hilbert (1899, 1913), Veblen (1904, 1911), Moore (1908) and Veblen and Young (1908). The first five axioms describe three-point properties of betweenness analogous to those of Euclidean and hyperbolic geometry.

Axiom O1
For events $a, b, c \in \mathcal{E}$,

$$[abc] \implies \exists Q \in \mathcal{P} : \; a, b, c \in Q \,.$$

Axiom O2
For events $a, b, c \in \mathcal{E}$,

$$[abc] \implies [cba] \,.$$

Axiom O3
For events $a, b, c \in \mathcal{E}$,

$$[abc] \implies a, b, c \;\; are \; distinct \,.$$

Axiom O4
For distinct events $a, b, c, d \in \mathcal{E}$,

$$[abc] \;\; and \;\; [bcd] \implies [abd] \,.$$

Axiom O5
For any path $Q \in \mathcal{P}$ and any three distinct events $a, b, c \in Q$,

$$[abc] \quad or \quad [bca] \quad or \quad [cab] \quad or$$
$$[cba] \quad or \quad [acb] \quad or \quad [bac] \quad .$$

The axiom of collinearity (Axiom O6) This axiom is a kinematic analogue of the geometric axiom of plane order given by Veblen (1904, 1911) and Moore (1908) and makes it possible to discuss "rectilinear motion" in terms of "collinear sets" of events and paths (as in Figure 1 which appears after the statement of the axiom). If a template is made by cutting a narrow slit in a sheet of paper, the paths may be observed "in motion" by moving the template gradually across the diagram.

A strict analogue of the geometric "Axiom of Pasch" (as stated by Veblen (1904) and Moore (1908))[5] would result in a non-independent system of axioms, so we state the corresponding Axiom O6 in terms of the concept of a "finite chain" in order to have an independent system of axioms. Accordingly we make the definition:
A sequence of events

$$Q_0, \ Q_1, \ Q_2, \ \cdots$$

(of a path Q) is called a *chain* if:
  (i)   it has two distinct events, or
  (ii) it has more than two distinct events and for all $i \geq 2$,

$$[Q_{i-2} \ Q_{i-1} \ Q_i] \ .$$

A *finite chain* is denoted by writing $[Q_0 \ Q_1 \ Q_2 \ \cdots \ Q_n]$ and an *infinite chain* is denoted by writing $[Q_0 \ Q_1 \ Q_2 \ \cdots]$ (note that the concept of a "chain" used by Veblen and Young (1908) for the discussion of projective geometries is entirely different from the concept defined above). Sometimes, for ease of reading, we will denote a chain or a relation of betweeness with commas to separate the events; thus, for example $[a, b, c]$ has the same meaning as $[abc]$. It will transpire (as a consequence of Theorem 1) that the concept of betweenness applies to any appropriately ordered triple of events of a finite chain, but note that this property is not being postulated in the axioms.

Axiom O6
If Q, R, S are distinct paths which meet at events $a \in Q \cap R$, $b \in Q \cap S$, $c \in R \cap S$ and if:
  (i)   *there is an event $d \in S$ such that $[bcd]$, and*
  (ii) *there is an event $e \in R$ and a path T which passes through both d and e such that $[cea]$,*

then T meets Q in an event f which belongs to a finite chain $[a \cdot \cdot f \cdot \cdot b]$.

As is the case with the analogous geometric axiom of plane order, this axiom makes statements about both incidence and order. The ensueing development of "order" properties on paths and properties of "collinear sets" has much in common with the treatment of order properties of points on lines and properties of planes as given by Veblen (1904, 1911). However the investigation of space–time is more complicated due to the existence of pairs of events which can not be connected by paths (see Axiom I5). As a consequence, the existence of "collinear sets" can only be established at the conclusion of Chapter 5 and is based upon thirty five preceding theorems.

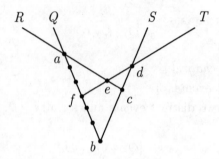

Figure 1   The configuration of Axiom O6

2.3 Axioms of incidence

Of the seven axioms of incidence which follow, the first four have counterparts in the axiomatic systems of Hilbert (1899), Veblen (1904, 1911) and Moore (1908) for Euclidean geometry. The fifth, sixth and seventh axioms describe properties of the "unreachable set" which distinguishes a space–time from a geometry. These latter properties are essentially properties of "non-incidence" and involve both order properties and incidence properties in their formulations. Axiom I5 excludes Galilean space–time as a possible model. The sixth and seventh incidence axioms (I6, I7) are stated in terms of the concept of a finite chain in order to establish the independence of the system of axioms; otherwise they could have been formulated more simply as in Theorems 13 and 4 (respectively) which describe the "connectedness" and the "boundedness" of the unreachable subset.

Axiom I1 (Existence)

$\mathcal{E}$ is not empty.

Axiom I2 (Connectedness)

For any two distinct events $a, b \in \mathcal{E}$ there are paths R, S such that $a \in R$, $b \in S$ and $R \cap S \neq \emptyset$.

Axiom I3 (Uniqueness)

For any two distinct events, there is at most one path which contains both of them.

In the subsequent development we will frequently be discussing the properties of sets of paths which meet at a given event. We will call any such set a *SPRAY of paths*, or more concisely a *SPRAY*, where the upper case letters indicate that we are referring to a set of paths rather than to a set of events: given any event x, we define

$$SPR[x] := \{R : R \ni x, \ R \in \mathcal{P}\} .$$

The corresponding set of events is called a *spray*, where the lower case letters indicate a set of events. We define

$$spr[x] := \{R_y : R_y \in R, \ R \in SPR[x]\} .$$

A subset of three paths $\{Q, R, S\}$ of a SPRAY is *dependent* if there is a path which does not belong to the SPRAY and which contains one event from each of the three paths: we also say that any one of the three paths is *dependent on* the other two. Otherwise the subset is *independent*.

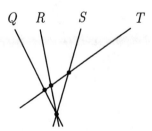

Figure 2   The subset $\{Q, R, S\}$ is dependent

We next give recursive definitions of dependence and independence which will be used to characterize the concept of dimension. A path T is *dependent on* the set of n paths (where $n \geq 3$)

$$\mathcal{S} = \{Q^{(i)} : \ i = 1, 2, \ldots, n; \ Q^{(i)} \in SPR[x]\}$$

if it is dependent on two paths $S^{(1)}$ and $S^{(2)}$, where each of these two paths is dependent on some subset of $n - 1$ paths from the set $\mathcal{S}$. We also say that the set of $n + 1$ paths $\mathcal{S} \cup \{T\}$ is a *dependent set*. If a set of paths has no dependent subset, we say that the set of paths is an *independent set*.

We now make the following definition which enables us to specify the dimension of Minkowski space–time. A SPRAY is a *3-SPRAY* if:

(i)   it contains four independent paths, and

(ii)   all paths of the SPRAY are dependent on these four paths.

Axiom I4 (Dimension)
If $\mathcal{E}$ is non–empty, then there is at least one 3-SPRAY.

Given a path Q and an event $b \notin Q$, we define the *unreachable subset of Q from b* to be

$$Q(b, \emptyset) := \{x : \text{ there is no path which contains } b \text{ and } x, \ x \in Q\} \ .$$

That is, the unreachable subset of Q from b is the subset of events of Q which can not be joined to b by a single path. If two events can not be connected by any path, we say that each is *unreachable* from the other; otherwise each is *reachable* from the other.

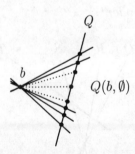

Figure 3   Pairs of events which can not be joined by a path will be indicated by a dotted line between them

Axiom I5 (Non–Galilean)

For any path Q and any event $b \notin Q$, the unreachable set $Q(b, \emptyset)$ contains (at least) two events.

Axiom I6 (Connectedness of the Unreachable Set)

Given any path Q, any event $b \notin Q$, and distinct events $Q_x, Q_z \in Q(b, \emptyset)$, there is a finite chain $[Q_0 \cdots Q_n]$ with $Q_0 = Q_x$ and $Q_n = Q_z$ such that for all $i \in \{1, 2, \ldots, n\}$,

$$\text{(i)} \quad Q_i \in Q(b, \emptyset)$$
$$\text{(ii)} \ [\, Q_{i-1} \, Q_y \, Q_i \,] \Longrightarrow Q_y \in Q(b, \emptyset) \ .$$

Axiom I7 (Boundedness of the Unreachable Set)

Given any path Q, any event $b \notin Q$ and events $Q_x \in Q \setminus Q(b, \emptyset)$ and $Q_y \in Q(b, \emptyset)$, there is a finite chain

$$[\, Q_0 \cdots Q_m \cdots Q_n \,]$$

with $\quad Q_0 = Q_x, \quad Q_m = Q_y \quad$ *and* $\quad Q_n \in Q \setminus Q(b, \emptyset) \ .$

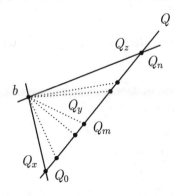

Figure 4   Boundedness of the Unreachable Set (Axiom I7)

2.4 Axiom of isotropy or symmetry

Compared to the absolute geometries, Minkowski space–time has the additional structure provided by the existence and properties of unreachable sets (Axioms I5, I6, I7). These properties, together with the single property of isotropy of the following axiom, are sufficient to take the place of all the axioms of congruence and the axiom of uniqueness of parallels used by Hilbert (1899), Moore (1908) and Veblen (1911) for Euclidean geometry.

For any two distinct paths Q, R which meet at an event x, we define the *unreachable subset of Q from Q_a via R* to be

$$Q(Q_a, R, x, \emptyset) := \{Q_y : [x \, Q_y \, Q_a] \text{ and } \exists R_w \in R \text{ such that } Q_a, Q_y \in Q(R_w, \emptyset) \} .$$

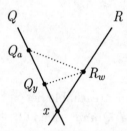

Figure 5   There is no path joining Q_a to R_w and there is no path joining Q_y to R_w

Axiom S (Symmetry or Isotropy)
If Q, R, S are distinct paths which meet at some event x and if $Q_a \in Q$ is an event distinct from x such that

$$Q(Q_a, R, x, \emptyset) = Q(Q_a, S, x, \emptyset)$$

then
  (i) *there is a mapping $\theta : \mathcal{E} \longrightarrow \mathcal{E}$*
  (ii) *which induces a bijection $\Theta : \mathcal{P} \longrightarrow \mathcal{P}$*
such that
  (iii) *the events of Q are invariant, and*
  (iv) $\Theta : R \longrightarrow S$.

The mapping θ is called a *symmetry mapping* or an *isotropy mapping*, with Q as the *invariant path*[6,7].

2.5 Axiom of continuity

This final axiom resembles the geometric axiom of the same name in the axiom systems of Hilbert (1899), Veblen (1904, 1911) and Moore (1908).

Given a path $Q \in \mathcal{P}$ and an infinite chain $[\, Q_0 \, Q_1 \, \cdots \,]$ of events in Q, the set

$$\mathcal{B} = \{Q_b : \; i < j \Rightarrow [Q_i \, Q_j \, Q_b]; \; Q_i, Q_j, Q_b \in Q\}$$

is called the *set of bounds* of the chain: if $\mathcal{B}$ is non–empty we say that the chain is *bounded*. If there is a bound $Q_b \in \mathcal{B}$ such that for all $Q_{b'} \in \mathcal{B} \setminus \{Q_b\}$,

$$[Q_0 \, Q_b \, Q_{b'}]$$

we say that Q_b is a *closest bound*.

Axiom C (Continuity)
Any bounded infinite chain has a closest bound.

3.   Temporal order on a path

In this chapter we will develop the properties of serial order of events on a path and then simplify and extend properties previously stated in some of the axioms of incidence, order and continuity. Many of the results of this chapter, especially those of Sections 3.5–3.7, are based on those of Veblen (1904, 1911) for affine and Euclidean geometry. Since there are pairs of events which can not be joined by a single path, the proofs are more complicated than those for the analogous geometric results.

3.1 Order on a finite chain

The properties of serial order of events are obtained in the later Section 3.6. In the present section we will obtain some preliminary results.

The Axiom of Uniqueness of Paths (Axiom I3) implies that *a path is uniquely determined by any two of its events:* accordingly if a path Q contains two distinct events a, b we may denote Q as ab (or as ba).

Theorem 1 *If $[abc]$ then $[cba]$ and no other order.*

Proof   (i) By Axiom O2, $[abc]$ implies $[cba]$. The other combinations would be (ii) $[bca]$ which by Axiom O2 is equivalent to $[acb]$, and (iii) $[cab]$ which by Axiom O2 is equivalent to $[bac]$. The case (ii) leads to a contradiction since, by Axiom O4, $[abc]$ and $[bca]$ imply $[aba]$ which contradicts Axiom O3. Similarly the case (iii) leads to a contradiction, since $[cab]$ and $[abc]$ imply $[cac]$.   *q.e.d.*

In the next theorem the relation of betweenness is extended from a set of three distinct events to the set of events of a chain.

Theorem 2 (Order on a Finite Chain)
On any finite chain $[Q_0 \cdots Q_n]$, there is a betweenness relation for each ordered triple; that is

$$0 \leq i < j < l \leq n \Longrightarrow [Q_i \, Q_j \, Q_l].$$

Furthermore all events of a chain are distinct.

Remark This result will be extended in Theorem 10 to any finite subset of events of a path.

Proof (i) We first show that

$$[abc] \quad \text{and} \quad [bcd] \Longrightarrow [acd] \tag{1}$$

Axiom O2 implies that $[dcb]$ and $[cba]$, so by Axiom O4 it follows that $[dca]$, whence Axiom O2 implies $[acd]$.

(ii) For convenience in this proof we will omit the symbol "Q" and simply indicate the suffices. The integers i, j, k, l will satisfy the order relations

$$0 \le i < j < n \quad \text{and} \quad 0 < k < l \le n$$

We define

$$P(i, j-1) := [i, \overline{j-1}, j]$$

Now for a finite chain of $n+1$ events we have $[\overline{j-1}, j, \overline{j+1}]$, so by part (i) above

$$P(i, j-1) \Longrightarrow P(i, j).$$

Furthermore, by definition of a chain, we have $P(i, i+1)$, so by induction we have, for all i and j,

$$i < j \Longrightarrow P(i, j) = [i, j, \overline{j+1}]. \tag{2}$$

Then as in the preceding paragraph, but with induction for a decreasing integer sequence, we have

$$k < l \Longrightarrow [\overline{k-1}, k, l]. \tag{3}$$

If we now let $k-1 = j$ then we have both $[i, j, j+1]$ and $[j, j+1, l]$ whence Axiom O4 implies $[i, j, l]$ for $i < j < l-1$. The case where $j = l-1$ has already been established in (2).

The first part of the theorem has now been established. Axiom O3 now implies that all events are distinct. *q.e.d.*

19

3.2 First collinearity theorem

A set of three distinct events $\{a, b, c\}$ is called a *kinematic triangle* if each pair of events belongs to one of three distinct paths: we will refer to the kinematic triangle $\triangle abc$, or simply $\triangle abc$.

Theorem 3 (Collinearity)
Given a kinematic triangle $\triangle abc$ and events d,e such that
   (i)   *there is a path de, and*
   (ii) *[bcd] and [cea]*

then de meets ab in an event f such that [afb].

Remark See Figure 1 (p.12).

Proof By the previous theorem (Theorem 2), the statement $[a \ldots f \ldots b]$ of the Axiom of Collinearity (Axiom O6) implies $[afb]$.   *q.e.d.*

3.3 Boundedness of the unreachable set

In this section we first show that each unreachable set is bounded in both directions. Then we obtain the Existence Theorem (Th.5) which will be used in subsequent proofs.

Theorem 4 (Boundedness of the Unreachable Set)
Let Q be any path and let b be any event such that $b \notin Q$. Given events $Q_x \in Q \setminus Q(b, \emptyset)$ and $Q_y \in Q(b, \emptyset)$, there is an event $Q_z \in Q \setminus Q(b, \emptyset)$ such that
   (i) $[Q_x \, Q_y \, Q_z]$,   *and*
   (ii) $Q_x \neq Q_z$.

Remark This theorem is illustrated by Figure 4 (p.15).

Proof This is an immediate consequence of Axiom I7 and Theorem 2.   *q.e.d.*

Theorem 5 (First Existence Theorem)
Given a path Q and an event $a \in Q$, there is
   (i)   *an event $b \in Q$ with b distinct from a, and*
   (ii) *an event $c \notin Q$ and a path ac (distinct from Q).*

Proof We first show that there is an event $d \notin Q$. Suppose the contrary; namely that each event is on the path Q: then the Axiom of Uniqueness of Paths (Axiom I3) implies that there is only one path, namely Q, which contradicts the Axiom of

Dimension (Axiom 1). Axiom I5 implies the existence of an event $b \in Q(d, \emptyset)$ with b distinct from a, which establishes (i).

The Axiom of Connectedness (Axiom I2) now implies the existence of a path R (distinct from Q) which meets Q at some event e. If this event is a then Axiom I5 implies the existence of an event c in $R(b, \emptyset)$ and the proof is complete; otherwise Axiom I5 and Theorem 4 imply the existence of a path (distinct from Q) through a which meets R at some event c. *q.e.d.*

3.4 Prolongation

Theorem 6 (Prolongation)
  (i) *If a, b are distinct events of a path Q, then there is an event $c \in Q$ such that $[abc]$.*
  (ii) *Each path contains an infinite set of distinct events.*

Proof (i) By the preceding theorem there is an event $e \notin Q$ and a path ae. By Axiom I5 there is an event $f \in ae(b, \emptyset)$ and so $b \in Q(f, \emptyset)$. Now Theorem 4 implies the existence of an event $c \in Q$ (with c distinct from a) such that $[abc]$.

(ii) By the preceding theorem any path Q has at least two distinct events. Now by part (i), Theorem 1, and induction, the path Q contains an infinite set of distinct events. *q.e.d.*

3.5 Second collinearity theorem

We will now extend the First Collinearity Theorem (Th.3) to the stronger and more useful Second Collinearity Theorem.

Theorem 7 (Second Collinearity Theorem)
In the notation of the axiom of collinearity (Axiom O6),

$$[afb] \quad and \quad [def].$$

That is, given a kinematic triangle $\triangle abc$ with $[bcd]$ and $[cea]$, if there is a path de, then on the path de there is an event f such that

$$[afb] \quad and \quad [def].$$

Remark See Figure 1 (p.12).

21

Proof (based on Veblen (1904), Theorem 7). The First Collinearity Theorem (Th.3) implies the existence of the event f such that $[afb]$: we now proceed to show that $[def]$. Since f is on the path de there are five possible orderings: $f = d$, $f = e$, $[efd]$, $[fde]$, or $[def]$. Either of the first two, by the Axiom of Uniqueness (Axiom I3), would imply the contradiction that a, b, c belong to the same path.

For the case $[efd]$, the First Collinearity Theorem (Th.3) applied to the kinematic triangle $\triangle dce$ with $[cea]$ and $[efd]$ (Figure 6a) implies the existence of an event x on af with $[dxc]$. By the Axiom of Uniqueness (Axiom I3) the paths af and cd can meet only once, so $x = b$ and $[dbc]$, which together with $[bcd]$ is a contradiction of Theorem 1.

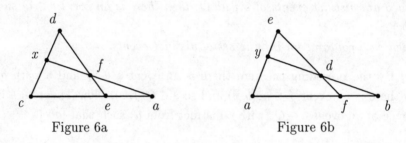

Figure 6a   Figure 6b

For the case $[fde]$, the First Collinearity Theorem (Th.3) applied to the kinematic triangle $\triangle eaf$ with $[afb]$ and $[fde]$ (Figure 6b) gives an event y on bd with $[eya]$. By the Axiom of Uniqueness, bd and ae can meet only once, so $y = c$ and $[eca]$, which together with $[cea]$, is a contradiction of Theorem 1.

The only remaining possibility is $[def]$.   *q.e.d.*

3.6 Order on a path

Before we establish the properties of serial order of events on a path (Theorems 9 and 10), we state and prove a preliminary result (Theorem 8) due to Veblen (1911, Theorem 6).

Theorem 8 (as in Veblen (1911) Theorem 6)
Given a kinematic triangle $\triangle abc$ with events a', b', c' such that $[ab'c]$, $[bc'a]$, and $[ca'b]$ then there is no path which contains a', b' and c'.

Proof (as in Veblen (1911) Theorem 6). By Axiom O3 the events a', b', c' are distinct from the events a, b, c. Now a', b', c' are distinct, since otherwise two

22

paths of the kinematic triangle $\triangle abc$ would meet at two distinct events which would contradict the Axiom of Uniqueness (Axiom I3).

We now suppose the contrary; namely, that there is a path which contains a', b', c' and we will proceed to obtain a contradiction. By Axiom O5 the events are in some order: first we consider the case $[a'b'c']$. There is no path which contains a', c', b because the Axiom of Uniqueness (Axiom I3) would then imply that this path should also contain the events a, c. Now the Second Collinearity Theorem (Th.7) implies that for the kinematic triangle $\triangle a'bc'$ with $[bc'a]$ and $[c'b'a']$ (by Theorem 1) the path ab' $(= ac)$ meets $a'b$ in an event x such that $[a'xb]$. The Axiom of Uniqueness (Axiom I3) implies that $c = ab' \cap a'b$ and so $x = c$, whence $[a'cb]$. Together with $[ca'b]$, this contradicts Theorem 1.

By cyclic interchange of the symbols a, b, c (and a', b', c') throughout the proof, the case $[b'a'c']$ also leads to a contradiction: the last case $[c'a'b']$ can be considered with a second cyclic interchange. $\hspace{2cm}$ *q.e.d.*

Theorem 9 *Any four distinct events on a path form a chain, so they may be represented by the symbols a, b, c, d in such a way that $[abcd]$.*

Proof The proof of this theorem is based on a theorem of Veblen (1904, Theorem 9, pp 356-357) and depends on Axiom O4 together with Lemmas 1, 2, and 3 which correspond (respectively) to Veblen's Lemmas 1, 2, 3, 5. In the present axiomatic system the proof of Lemma 1 (which corresponds to Veblen's Lemma 2) is modified by inserting the first four paragraphs. Veblen's Lemmas 4 and 6 are not required here because we already have established the result of Theorem 2 for the order of events on a finite chain.

Lemma 1 *If $[abc]$ and $[abd]$ and $c \neq d$ then either $[bcd]$ or $[bdc]$.*

Proof (based on Veblen (1904), Lemma 2, p.357). By Theorem 1, it is sufficient to show that the supposition $[dbc]$ leads to a contradiction.

By the Existence Theorem 5 there is a path S distinct from ab passing through a (see Figure 7, p.24). Axiom I5 together with Theorem 4 imply the existence of an event $e \in S \setminus \{a\}$ and a path be. If there is a path de we let $d^* := d$. Otherwise the Boundedness of the Unreachable Set (Th.4) implies the existence of both an event $d^* \in ab$ and a path d^*e such that $[bdd^*]$, which with $[abd]$ implies $[abd^*]$ by Axiom O4.

Similarly if there is a path ce we let $c^* := c$. Otherwise there is an event $c^* \in ab$ and a path c^*e such that $[bcc^*]$ which with $[abc]$ (and $[dbc]$) implies $[abc^*]$ (and $[dbc^*]$).

By Theorem 1 $[abc^*]$ implies $[c^*ba]$ and $[dbc^*]$ implies $[c^*bd]$ so, in the case where $d^* \neq d$ we have $[bdd^*]$ so Axiom O4 implies $[c^*bd^*]$, and in the case where $d^* = d$, we also have $[c^*bd^*]$ (from $[c^*bd]$).

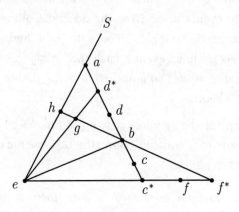

Figure 7

By the Prolongation Theorem (Th.6) there is an event $f \in ec^*$ such that $[ec^*f]$. If there is a path bf we let $f^* := f$. Otherwise the Boundedness of the Unreachable Set (Theorem 4) implies the existence of an event $f^* \in ec^*$ and a path bf^* such that $[c^*ff^*]$, which with $[ec^*f]$ implies $[ec^*f^*]$ by Axiom O4. Thus for both cases there is an event $f^* \in ec^*$ and a path bf^* such that $[ec^*f^*]$.

The remainder of this proof is based on a proof of Veblen (Veblen (1904), Lemma 2, p357, with the symbols e, f^*, g, h taking the place of Veblen's symbols O, P, Q, R respectively). For the kinematic triangle $\triangle aec^*$ with $[ec^*f^*]$ and $[c^*ba]$, the Second Collinearity Theorem (Th.7) implies that f^*b meets ae in an event h such that $[ahe]$ and $[f^*bh]$, while the same theorem applied to the kinematic triangle $\triangle d^*ec^*$ with $[ec^*f^*]$ and $[c^*bd^*]$ implies that f^*b meets d^*e in an event g such that $[d^*ge]$ and $[f^*bg]$. Since both g and h lie on f^*b, it follows that b, g, h belong to the same path: however b, g, h are events of segments of the kinematic triangle $\triangle ead^*$ since $[abd^*]$, $[d^*ge]$ and $[ahe]$. This contradicts the result of Theorem 8, and completes the proof of Lemma 1.

Lemma 2 *If $[abc]$ and $[abd]$ and $c \neq d$ then either $[acd]$ or $[adc]$.*

Proof By Lemma 1, either $[bcd]$ or $[bdc]$. Thus by Theorem 1, we have both $[cba]$ and $[dba]$ as well as either $[dcb]$ or $[cdb]$, so by Axiom O4,

$$[dcb] \quad \text{and} \quad [cba] \Longrightarrow [dca]$$
$$[cdb] \quad \text{and} \quad [dba] \Longrightarrow [cda]$$

and then Theorem 1 implies $[acd]$ or $[adc]$.

Lemma 3 *If $[abc]$ and $[acd]$ then $[bcd]$.*

Proof We suppose the contrary, namely that $[bdc]$ or $[cbd]$ and will obtain contradictions in both cases. In the first case, Theorem 1 implies $[cdb]$ and then Axiom O4 together with $[acd]$ implies $[acb]$ which contradicts Theorem 1. In the second case, the same theorem implies $[cba]$ and then Lemma 2 with $[cbd]$ and $a \neq d$ implies $[cad]$ or $[cda]$, which both contradict $[acd]$. This completes the proof of Lemma 3.

The proof of Theorem 9 can now be completed. By Axiom O5 we can take any three of the four distinct events and assign the symbols a^*, b^*, c^* so that $[a^*b^*c^*]$. Again by Axiom O5 the fourth event d^* will then satisfy either:

(i) $[d^*a^*b^*]$ so the four events form the chain $[d^*a^*b^*c^*]$
(ii) $[a^*d^*b^*]$, whence Lemma 3 implies $[d^*b^*c^*]$ and so $[a^*d^*b^*c^*]$
(iii) $[a^*b^*d^*]$, whence Lemma 1 implies either $[b^*c^*d^*]$ or $[b^*d^*c^*]$. In the first case where $[b^*c^*d^*]$, the events a^*, b^*, c^*, d^* form the chain $[a^*b^*c^*d^*]$ while in the second case where $[b^*d^*c^*]$, Lemma 3 and Theorem 1 imply that the events form the chain $[a^*b^*d^*c^*]$.

For each case the events can now be relabelled from a^*, b^*, c^*, d^* to a, b, c, d in some order so that $[abcd]$. $\qquad\qquad$ *q.e.d.*

The next theorem is a converse result to Theorem 2 and is an explicit statement of the properties of linear (or serial) order of events on a path.

Theorem 10 (based on Veblen (1904), Theorem 10)
Any finite set of distinct events on a path form a chain. That is, any set of n distinct events can be represented by the notation $a_1, a_2, \ldots, a_n$ such that

$$[a_1 a_2 \cdots a_n].$$

Remark Thus the notation [$\cdots$] denotes an ordering which is irreflexive and transitive and which could be represented as a linear ordering (as will be done before the statement of Theorem 29).

Proof (By induction). The previous theorem applies to the case where $n = 4$. We will make the inductive hypothesis that the result applies to a set of n distinct events $\{a_1, a_2, \ldots, a_n\}$ and demonstrate that this implies the result for the case of $n + 1$ distinct events. We denote the $(n + 1)$–th event as b. Then Axiom O5 implies that either:

(i) $[ba_1a_n]$ or (ii) $[a_1ba_n]$ or (iii) $[a_1a_nb]$

Case (i): By the inductive hypothesis and Theorem 2 we have $[a_1a_2a_n]$ so the previous theorem (Th.9) implies that $[ba_1a_2a_n]$ which implies that $[ba_1a_2]$. Thus b is an element of a chain $[a_1^*a_2^*\ldots a_{n+1}^*]$ where $a_1^* := b$ and (for $j \in \{2, \ldots, n+1\}$) $a_j^* := a_{j-1}$.

Case (ii): Let k be the smallest integer such that $[a_1ba_k]$. Then the previous theorem (Th.9) implies either that $[a_1a_{k-1}ba_k]$, or that $k = 2$ so that $[a_{k-1}ba_k]$. If $k - 2 \geq 1$ we have $[a_{k-2}a_{k-1}a_k]$ which with $[a_{k-1}ba_k]$ implies $[a_{k-2}a_{k-1}ba_k]$ by the previous theorem, while if $k + 1 \leq n$ we have $[a_{k-1}a_ka_{k+1}]$ which with $[a_{k-1}ba_k]$ implies $[a_{k-1}ba_ka_{k+1}]$; that is we have now shown that $[a_{k-2}a_{k-1}b]$ (if $k - 2 \geq 1$) and $[a_{k-1}ba_k]$ and $[ba_ka_{k+1}]$ (if $k + 1 \leq n$) so that b is an element of a chain $[a_1^*a_2^*\ldots a_{n+1}^*]$ where

$$a_j^* = \begin{cases} a_j, & j \leq k - 1 \\ b, & j = k \\ a_{j-1}, & j > k . \end{cases}$$

Case (iii): The proof for this case is similar to that for Case (i).

Thus in each case the inductive hypothesis for the case of n distinct events implies the result for the case of $n + 1$ distinct events and the result has been established in the previous theorem for the case $n = 4$: the proof is now complete.   *q.e.d.*

Given any two distinct events a, b of a path we define the *segment*

$$(ab) := \{ x : [axb], \ x \in ab \}$$

and the *interval*

$$|ab| := (ab) \cup \{a, b\}$$

where the interval is the union of the corresponding segment with its end-points.

The set of events

$$\{\, x : [abx] : \; x \in ab \,\}$$

is called the *prolongation of the segment* (ab) *beyond* b. The segment (ab) together with the singleton set $\{a\}$ and the prolongation beyond b is called the *ray* $\overrightarrow{ab}$.

Theorem 11 (after Veblen (1904), Theorem 11)
Any finite set of N distinct events of a path separates it into $N-1$ segments and two prolongations of segments.

Proof   As in the proof of the previous theorem (Th.10), any event distinct from the a_i $(i = 1, \ldots, N)$ belongs to a segment (Case (ii)) or a prolongation (Cases (i) and (iii)). Theorem 1 implies that the $N-1$ segments and two prolongations are disjoint.
$$q.e.d.$$

3.7   Continuity and the monotonic sequence property

It is convenient to have a notation for *non-strict ordering*, so we define

$$[ab \ldots ef]\!] := [ab \ldots ef] \quad \text{or} \quad e = f$$
$$[\![ab \ldots ef] := [ab \ldots ef] \quad \text{or} \quad a = b$$
$$[\![ab \ldots ef]\!] := [\![ab \ldots ef] \quad \text{or} \quad [ab \ldots ef]\!]$$

where the non-strictness applies only to the end events.

Theorem 12 (Continuity)

(i)   (Two Rays). (Based on Veblen (1911) Theorem 9). *Given a path Q and an event $a \in Q$, the events of Q exclusive of a belong to two rays such that a is between any event of one ray and any event of the other ray, but a is not between any two events of the same ray.*

(ii)   (Continuity). *Given a path Q and a partition of the set of events of the path into two non–empty sets $\mathcal{L}$ and $\mathcal{R}$ such that no event of either set is between any two events of the other set and $\mathcal{L}$ has no event closest to $\mathcal{R}$, then there is an event b which belongs to $\mathcal{R}$ and is between each element of $\mathcal{L}$ and each other element of $\mathcal{R}$.*

Remark   Similar results apply to a segment, or an interval, of a path.

Proof (i) The Prolongation Theorem (Th.6) implies the existence of an event b of Q distinct from a. Axiom O5 and Theorem 1 imply that

$$Q = \{\, x : [xab]\,\} \cup \{\, a\,\} \cup \{\, y : [ayb] \text{ or } [aby]\,\}$$

where the two sets of events not containing a are disjoint and will be referred to as $\mathcal{L}$ and $\mathcal{R}$, respectively. Then Theorem 10 and Theorem 2 imply that for distinct $w, x \in \mathcal{L}$ and for distinct $y, z \in \mathcal{R}$

$$[xayb] \text{ or } [xaby],$$
$$[wxay] \text{ or } [xway], \text{ and}$$
$$[xayz] \text{ or } [xazy].$$

(ii) Take any events $l \in \mathcal{L}$ and $r \in \mathcal{R}$ and let $\mathcal{L}^* := \mathcal{L} \cap |lr|$ and let $\mathcal{R}^* := \mathcal{R} \cap |lr|$. Consider an infinite chain

$$\mathcal{C}^{(1)} = [\, l, l_1^{(1)}, l_2^{(1)}, l_3^{(1)}, \cdots] \qquad \text{with} \quad l_i^{(1)} \in \mathcal{L}^*.$$

Since $\mathcal{C}^{(1)}$ is bounded by r, its set of bounds $\mathcal{B}^{(1)}$ contains r and is non–empty: the Axiom of Continuity (Axiom C) implies that $\mathcal{B}^{(1)}$ has a closest bound $b^{(1)}$ (which could be the same event as r) so $\mathcal{B}^{(1)} = |b^{(1)}r|$ (which is equal to the singleton set $\{r\}$ in the case where $b^{(1)} = r$). Similar considerations apply to each infinite chain

$$\mathcal{C}^{(\lambda)} = [\, l, l_1^{(\lambda)}, l_2^{(\lambda)}, l_3^{(\lambda)}, \cdots]$$

which therefore has a closest bound $b^{(\lambda)}$ and a set of bounds $\mathcal{B}^{(\lambda)} = |b^{(\lambda)}r|$. The intersection of all the $\mathcal{B}^{(\lambda)}$ for all infinite chains $\mathcal{C}^{(\lambda)}$ is the set of all events which bound every infinite chain $\mathcal{C}^{(\lambda)} \in \mathcal{L}^*$. Now each $\mathcal{B}^{(\lambda)}$ is a closed interval containing r (or a singleton set containing r) so the intersection of all the $\mathcal{B}^{(\lambda)}$ is a closed interval (or a singleton set containing r). If the closed set is an interval, it has another end-point b distinct from r while if the closed set is a singleton set we let $b := r$. In either case, b is a bound for every chain $\mathcal{C}^{(\lambda)} \in \mathcal{L}^*$.

Since $\mathcal{L}^*$ has no event closest to $\mathcal{R}$, the bound b can not belong to $\mathcal{L}^*$ (since otherwise it would be an event of some chain $\mathcal{C}^{(\mu)}$): thus $b \in \mathcal{R}^*$. There is no event $b' \in \mathcal{R}^*$ such that $[lb'br]$, since such an event b' would belong to each set of bounds $\mathcal{B}^{(\lambda)}$ and so b' would have to be closer (to $\mathcal{L}^*$) than the closest bound b. Thus b is

28

the closest event of $\mathcal{R}^*$ (to $\mathcal{L}^*$) and hence

$$\mathcal{L}^* = \{l\} \cup (lb) \qquad \text{and} \qquad \mathcal{R}^* = |br|$$

which completes the analogue of (ii) for the restriction to the segment $|lr|$. The extension to the set of all events of Q is an obvious consequence of (i). $\qquad$ *q.e.d.*

3.8 Connectedness of the unreachable set

In Section 3.3 we showed that each unreachable set is bounded in both directions.

We next show that unreachable sets are connected with respect to the usual order topology on each path. Then we obtain a more useful form of existence theorem.

Theorem 13 (Connectedness of the Unreachable Set)
Given any path Q, any event $b \notin Q$, and distinct events $Q_x, Q_z \in Q(b, \emptyset)$, then

$$[Q_x \, Q_y \, Q_z] \Longrightarrow Q_y \in Q(b, \emptyset).$$

Remark Thus $Q(b, \emptyset)$ is connected and it separates $Q \backslash Q(b, \emptyset)$ into two components: each of these two components is connected.

Proof By Axiom I6 there is a finite chain $[Q_0 \, Q_1 \dots Q_{n-1} \, Q_n]$ (where $Q_0 = Q_x$ and $Q_n = Q_z$) so Theorem 11 implies that for some $i \in \{1, \; \dots, n\}$, $[Q_{i-1} \, Q_y \, Q_i]$ whence Axiom I6 implies that $Q_y \in Q(b, \emptyset)$. $\qquad$ *q.e.d.*

The notations for *betweenness relations* are extended to apply to sets of events if the relations are satisfied for all cases of individual events from the sets; thus, for example we write

$$[x \, \mathcal{A} \, z] \quad \Longleftrightarrow \quad \text{for all } y \in \mathcal{A} \,, \; [xyz]$$
$$[x \, \mathcal{A} \mathcal{B}] \quad \Longleftrightarrow \quad \text{for all } y \in \mathcal{A} \,, \text{for all } z \in \mathcal{B} \,, \; [xyz] \,.$$

In order to facilitate the reading of betweenness relations, commas will sometimes be used; so for example

$$[y, Q(a, \emptyset), z] \Longleftrightarrow [y \, Q(a, \emptyset) \, z] \,.$$

29

Theorem 14 (Second Existence Theorem)

(i)   *Given a path Q and a pair of events $a, b \notin Q$ each of which can be joined to Q by some path, there are events $y, z \in Q$ such that*

$$[y, Q(a, \emptyset), z] \qquad and \qquad [y, Q(b, \emptyset), z].$$

(ii)   *Given a path Q and a pair of events $a, b \notin Q$ each of which can be joined to Q by some path and a pair of events $c, d \in Q$, there is an event $e \in Q$ and paths ae, be such that $[cde]$.*

(iii)   *Given two paths Q and R which meet at x, an event $a \in R \setminus \{x\}$ and an event $b \notin Q$ which can be joined to Q by some path, there is an event e and paths ae, be such that $[x, Q(a, \emptyset), e]$.*

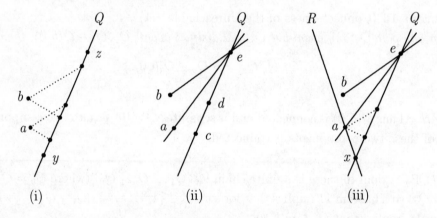

(i)   (ii)   (iii)

Figure 8   The Second Existence Theorem

Proof   (i) Theorem 4 implies that both sets $Q(a, \emptyset)$ and $Q(b, \emptyset)$ are bounded in both directions by events which do not belong to the unreachable sets themselves, so the union $Q(a, \emptyset) \cup Q(b, \emptyset)$ is bounded by distinct events y, z which do not belong to the union of the unreachable sets.

(ii) In the cases where $[cdy]$ or $[cdz]$ we define e to be y or z respectively. The other cases are where $(\llbracket ycd \rrbracket$ or $[cyd \rrbracket)$ and $(\llbracket zcd \rrbracket$ or $[czd \rrbracket)$: in these cases the Prolongation Theorem (Th.6) implies the existence of an event e such that $[cde]$ and by Theorem 10 the event e is not between the bounding events y, z so there are paths ae, be.

(iii) By (ii) above, if we let $c := x$ and take any $d \in Q(a, \emptyset)$ there is an event $e \in Q$ and paths ae, be such that $[xde]$. Theorem 13 then implies that $[x, Q(a, \emptyset), e]$.

q.e.d.

4. Collinearity and temporal order

In this chapter we extend the previous collinearity properties and are able to define and deduce useful properties of "compact collinear sets" in Section 4.4. In Section 4.5 we deduce properties of "signal" and "record" functions which lead to the important Causality Theorem 26 of Section 4.6.

4.1 Third collinearity theorem

Given a kinematic triangle $\triangle a_1 a_2 a_3$, the three intervals $|a_1 a_2|$, $|a_2 a_3|$, $|a_3 a_1|$ are called *sides* and their union is called the *boundary* $\mathcal{B}(a_1, a_2, a_3)$. Any event between two events of distinct sides of the triangle is called an *internal event*.

A path which meets the boundary at exactly two distinct events separates the remaining subset of boundary events into two components: if a second path meets each of these components at exactly one event, we say that the two paths *cross each other over the kinematic triangle*.

Theorem 15 (Third Collinearity Theorem)

(i)   *If abc and dbf are two kinematic triangles such that* $[afb]$ *and* $[bcd]$, *then* (ac) *meets* (df) *in an event e (see Figure 9).*

(ii) *If two paths cross each other over a kinematic triangle, then they meet at an internal event.*

(iii) *Two paths which meet at an internal event (and which meet the boundary at four distinct events), cross each other over the kinematic triangle.*

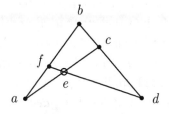

Figure 9

Remarks Part (i) is a converse result to the Second Collinearity Theorem (Th.7). We will first prove Lemma 1 and then part (i). Next we will prove Lemma 2 followed by parts (ii) and (iii).

Lemma 1 *For any events a, b, c, d, f such that there are paths ab, ac, ad, bc, df and*

$$[bcd] \,, \quad a \notin bc \,, \quad [afb]$$

there is an event e such that $[aec]$ and $[def]$ (see Figure 10a).

Remarks This lemma is almost a converse of Theorem 7 except that here we postulate the existence of the path ad. Lemma 1 will be superseded by (i) (of this Theorem 15) which is converse to Theorem 7.

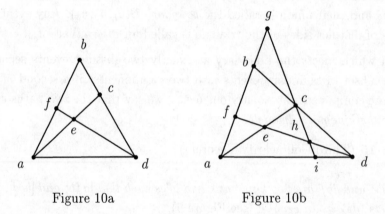

Figure 10a   Figure 10b

Proof (of Lemma 1 which is based on Veblen (1904), Theorem 13, p358).
The preceding theorem (Th.14(ii)) implies the existence of an event g and a path cg such that $[afbg]$ (as in Figure 10b). Now the Second Collinearity Theorem (Th.7) applied to the kinematic triangle $\triangle dfb$ with $[fbg]$ and $[bcd]$ implies the existence of an event h such that $[dhf]$ and $[gch]$. Similarly for the kinematic triangle $\triangle daf$ with $[afg]$ and $[fhd]$ there is an event i such that $[dia]$ and $[ghi]$, and then Theorem 10 implies $[gchi]$. Another application of the Second Collinearity Theorem to the kinematic triangle $\triangle cai$ with $[aid]$ and $[ihc]$ implies the existence of an event e such that $[cea]$ and $[dhe]$. A final application of the same theorem to the kinematic triangle $\triangle abc$ with $[bcd]$ and $[cea]$ implies that ($[aec]$ and) $[def]$ which completes the proof of Lemma 1.

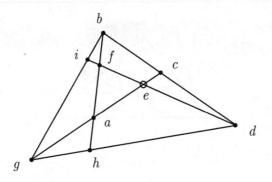

Figure 11

Proof (of (i)). The Second Existence Theorem (Th.14) implies the existence of an event g and paths gb, gd such that $[gac]$ (Figure 11). Then the Second Collinearity Theorem (Th.7) applied to the kinematic triangle $\triangle gdc$ with $[dcb]$ and $[cag]$ implies the existence of an event h such that $[ghd]$ and $[bah]$, whence Theorem 10 implies $[bfah]$. Again, the Second Collinearity Theorem applied to the triangle $\triangle bgh$ with $[ghd]$ and $[hfb]$ implies that df meets bg in an event i such that $[big]$. Finally Lemma 1 (above) (with g, b, c, d, i taking the place of a, b, c, d, f respectively) implies the existence of an event e such that $[gec]$ and $[dei]$.

Now Theorem 10 with $[gac]$ and $[gec]$ implies that either: (α) $a = e$, (β) $[geac]$ or (γ) $[gaec]$. Case (α) would imply that $a(= e) = f$ which would contradict $[bfah]$. Case (β) with the Second Collinearity Theorem applied to the kinematic triangle $\triangle gha$ with $[haf]$ and $[aeg]$ gives $[gdh]$ which contradicts the previously obtained $[ghd]$. The only remaining possibility is Case (γ) which implies $[aec]$. Since the statement of the theorem is symmetric with respect to interchange of the symbols a with d and f with c, the second order relation $[def]$ can be established in a similar manner: this completes the proof of (i).

Lemma 2 *Let $b_1 b_2 b_3$ be a kinematic triangle and let ac, df be paths such that $[b_1 f b_2]$, $[b_2 a d b_3]$, $[b_1 c b_3]$. Then ac meets df in an event e such that $[aec]$ and $[def]$ (see Figure 12).*

Proof Case (a): If fd meets $b_1 b_3$ in an event g such that $[b_1 b_3 g]$ (Figure 13a), then the Second Collinearity Theorem (Th.7) applied to the kinematic triangle $\triangle acb_3$ with $[(b_1) cb_3 g]$ and $[b_3 da]$ implies the existence of an event e such that $[aec]$ and $[gde]$, and the same theorem applied to the kinematic triangle $\triangle b_2 b_1 b_3$ with $[b_1 b_3 g]$

33

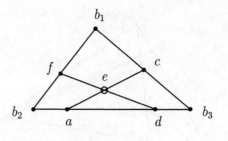

Figure 12

and $[b_3db_2]$ implies that $[gdf]$.

Theorem 8 implies that e and f are distinct so either (α) $[gdef]$ or (β) $[gdfe]$: we will show that (α) is true by demonstrating that (β) leads to a contradiction. By part (i), $[dfe]$ implies that the interval $|b_2f|$ ($\subseteq |b_2b_1|$) meets the interval $|ae|$ ($\subseteq ac$) in an event i such that $[aiec]$. Part (i) applied to the configuration $egcb_1f$ (which has $[efg]$ and $[gcb_1]$) implies that the segments (ec) and (b_1f) meet at an event j, but these segments belong to the paths ac and b_2b_1 respectively, so the paths meet at the two distinct events i and j: this contradicts the Axiom of Uniqueness (Axiom I3). We have now shown for Case (a) that $[aec]$ and $[def]$.

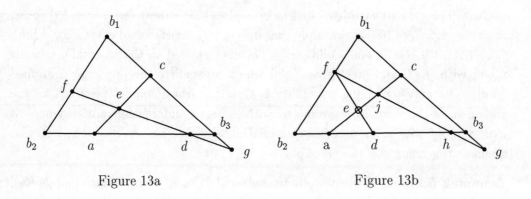

Figure 13a   Figure 13b

Case (b): If fd does not meet b_1b_3 in an event g such that $[b_1b_3g]$ (Figure 13b), then the Second Existence Theorem (Th.14) implies that there is some event g and a path fg such that $[b_1b_3g]$. Part (i) above implies that fg meets b_2b_3 in an event h such that $[fhg]$ and $[b_2hb_3]$. Now for this case, h and d are distinct so either (α) $[b_2dhb_3]$ or (β) $[b_2hdb_3]$, but (β) leads to a contradiction since the Second Collinearity Theorem applied to the kinematic triangle $\triangle b_3gh$ with $[ghf]$ and $[hdb_3]$ would imply that fd

meets b_1b_3 in an event g' (such that $[(b_1)\, b_3\, g'g]$). Thus (α) applies, so $[b_2dhb_3]$ and by Case (a), ac meets fh at an event j such that $[ajc]$, and $[hjf]$ so by part (i) fd meets $ac\ (= aj)$ at an event e such that $[aej(c)]$ and $[def]$. This completes the proof of Case (b) and hence the proof of Lemma 2.

The proof of part (ii) of Theorem 15 is now complete because all possible configurations occur as special cases of part (i) and Lemma 2.

Part (iii) is an immediate consequence of part (ii), the Second Collinearity Theorem (Th.7) and Theorem 8. $\hfill q.e.d.$

4.2 There is no fastest path

Theorem 16 (There is no fastest path)
Let Q be a path and let e be an event not in Q.
  (i) *For each event $c \in Q \setminus Q(e,\emptyset)$ there is some event $b \in Q \setminus Q(e,\emptyset)$ such that $[Q(e,\emptyset), b, c]$.*
  (ii) *Each component of $Q \setminus Q(e,\emptyset)$ is open with respect to the order topology.*
  (iii) *$Q(e,\emptyset)$ is a closed interval.*

Proof Since the unreachable set $Q(e,\emptyset)$ is bounded (Theorem 4) and connected (Theorem 13), $Q \setminus Q(e,\emptyset)$ has two components.

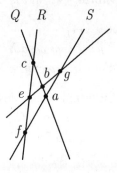

Figure 14

(i) Let $R := ec$ and consider a fixed event $a \in Q(e,\emptyset)$ (Figure 14). Since $e \in R(a,\emptyset)$, Theorem 4 implies that there is an event $f \in R$ and a path $S\ (= af)$ such that $[cef]$. Similarly $a \in S(e,\emptyset)$ so there is an event g and a path eg such that $[fag]$. Now the preceding theorem (Th.15(i)) implies that eg meets ac in an event b

such that $[abc]$. Since the unreachable set is connected, the order $[abc]$ applies for all $a \in Q(e, \emptyset)$.

(ii) By Theorems 4 and 13, the set $Q \setminus Q(e, \emptyset)$ consists of two components, each of which is connected. By part (i), each component is open.

(iii) The Continuity Theorem (Th.12) implies that $Q(e, \emptyset)$ is closed and, since it is connected, it is an interval.   *q.e.d.*

4.3 Each path is dense in itself

Theorem 17 (Each path is dense in itself)
Given any path Q with distinct events $a, c \in Q$,
there is an event $b \in Q$ such that $[abc]$.

By the First Existence Theorem (Th.5) there is an event $d \notin Q$ and a path cd. Axiom I5 implies the existence of an event $e \in cd(a, \emptyset)$. Now the conditions of the preceding theorem (Th.16) are satisfied so there is an event $b \in Q$ such that $[abc]$.
                                                                                    q.e.d.

4.4 Compact collinear set

Given a kinematic triangle $\triangle abc$, the union of the set of internal events with the boundary is called a *compact collinear set* and is denoted by $\mathcal{C}(a, b, c)$.

Theorem 18 (Compact Collinear Set)
If two events of $\mathcal{C}(a, b, c)$ can be joined by a path, then the path meets the boundary in two events and all events of the corresponding interval belong to $\mathcal{C}(a, b, c)$.

Proof We first show that any internal event lies on a path which joins a vertex to a side. Consider an event f such that $[gfh]$ with $[agb]$ and $[bhc]$ (Figure 15). Theorem 13 implies that either there is a path fc or there is a path fb. In the case where there is a path fc, Theorem 3 implies that cf meets ab; while for the case where there is a path fb, Theorem 15 implies the existence of an event i and a path ci such that $[fib]$, whence Theorem 15(i) implies the existence of an event j such that $[cji]$ and $[fjh]$. Now the Second Collinearity Theorem (Th.7) implies that cj meets ab in an event k such that $[agkb]$ and a second application of the same theorem shows that bi $(= bf)$ meets (ac) at an event l. We have now shown that any internal event lies on a path which joins a vertex to an event of the opposite side.

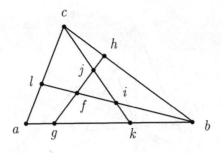

Figure 15

Theorem 13 now implies that there is at most one vertex which can not be reached from a given internal event, so for any two given internal events d, e there is at least one vertex from which paths join d, e (respectively) to the opposite side. Without loss of generality we shall relabel vertices so that this vertex is denoted as c and we shall define

$$k := cd \cap ab \quad \text{and} \quad l := ce \cap ab.$$

such that $[aklb]$ or $k = l$. The case $k = l$ is trivial so we proceed to consider the case $[aklb]$. The Second Existence Theorem (Th.14) implies the existence of an event m and a path dm such that $[bam]$ and so by the Third Collinearity Theorem (Th.15(i)) there is an event n such that $[mnd]$ and $[anc]$. Also the Second Collinearity Theorem (Th.7) implies the existence of an event p such that $[cpl]$ and $[m(n)dp]$. We will demonstrate that for all possible cases where there is a path de there is a boundary event h such that $[hde]$. There are four possible cases (which are illustrated in Figure 16 by e, p, e', e^* respectively).

Case (i): If $[cpel]$ the Second Collinearity Theorem implies the existence of an event h such that $[(a)nhc]$ and $[edh]$.

Case (ii): If $e = p$ then we let $h := n$.

Case (iii): If $[ce'pl]$ then the Second Collinearity Theorem implies that $e'd$ meets ml in an event h' such that $[mh'l]$ and $[e'dh']$. Now if $h' = a$ or if $[ah'l(b)]$ there is nothing further to prove while

Case (iii*): If $[ce^*pl]$ and $[(m)h^*al]$ the Third Collinearity Theorem implies the existence of an event $\overline{h}$ such that $[a\overline{h}nc]$ and $[h^*\overline{h}d(e^*)]$.

37

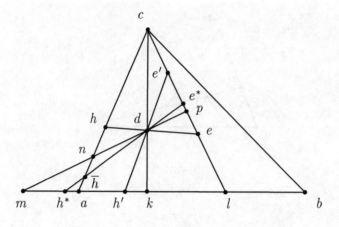

Figure 16

Thus in all cases there is a boundary event h (respectively n, h', $\overline{h}$ for cases (ii), (iii), (iii*)) such that $[hde]$. Similarly there is a boundary event f such that $[def]$, which completes the proof. *q.e.d.*

4.5 Optical line and record function theorems

In this section we define the relation of events being "in optical line" and then establish the First Optical Line Theorem (Th.19) and its corollary (Th.20). We then show in Theorems 21 and 22 that isotropy mappings are bijections from $\mathcal{E}$ to $\mathcal{E}$ and that they preserve some order properties. In Theorems 24 and 25 we obtain properties of "record functions" and "reverse record functions" which will be used in the following Section 4.6 to establish the Causality Theorem (Th.26).

Given a path Q, a subset $\mathcal{S} \subset Q$ and events $a \in Q \setminus \mathcal{S}$ and $b, c \in \mathcal{S}$ we say that:

(i) b is the *closest event of* $\mathcal{S}$ *to* a if $[ab\mathcal{S}]$, and

(ii) c is the *furthest event of* $\mathcal{S}$ *from* a if $[a\mathcal{S}c]$.

Theorem 16 implies that unreachable sets are intervals whose extreme events or end–events may be conveniently specified using the terms "closest event to" and "furthest event from".

Theorem 19 (First Optical Line Theorem)

Let abd be a kinematic triangle and let c be an event such that [bcd] and such that there is a path ac.

(i) *The unreachable set $ab(d, \emptyset)$ is between a and b if and only if the unreachable set $ac(d, \emptyset)$ is between a and c.*

For an unreachable set $ab(d, \emptyset)$ between a and b, let f (resp. f') be the closest (resp. furthest) event to (resp. from) a and similarly let e (resp. e') be the closest (resp. furthest) event of $ac(d, \emptyset)$ to (resp. from) a (see Figure 17).

(i') *Paths which meet ab at events between a and f (resp. f' and b) meet ac at events between a and e, (resp. e' and c), and conversely.*

(ii) *f is the event of $ab(e, \emptyset)$ closest to a.*

(ii') *f' is the event of $ab(e', \emptyset)$ furthest from a.*

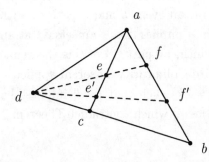

Figure 17   We adopt the convention that a dashed line indicates the relation between an event and an end–event of its unreachable set

Remarks   1. The relationship between the three events d, e, f of (part (ii)) will be described by saying that d, e, f are *in optical line*. Similarly we will say that d, e', f' (of part (ii')) are *in optical line*.

2. At this stage we are neither assuming nor proving that the relationships between an event and an end–event of its unreachable set are symmetric.

Proof (i) The Second and Third Collinearity Theorems (Th.7 and Th.15) imply that the set of paths which join d to events of the segment (ab) define a strictly order-preserving bijection between $(ab) \setminus ab(d, \emptyset)$ and $(ac) \setminus ac(d, \emptyset)$.

If $ab(d, \emptyset)$ is between a and c then by Theorem 16, $(ab) \setminus ab(d, \emptyset)$ consists of two open components, so each of these components is mapped bijectively onto an open component of $(ac) \setminus ac(d, \emptyset)$. Furthermore the Continuity Theorem (Th.12) implies that there is at least one event between these components and so by Theorem 13, $ac(d, \emptyset)$ is between a and c. The converse proposition is established in a similar manner (simply interchange the symbols b, c).

(i′) The ordering property stated in (i′) is an immediate consequence of the order-preserving bijection described in the first paragraph (of the proof of (i) above).

(ii) We need to show firstly that there is no path from e to f and secondly that f is the closest event of $ab(e, \emptyset)$ to a. We will show that there is no path from e to f by supposing the contrary and obtaining a contradiction. The supposition that ef exists implies, by Theorem 18 and the definition of f, that ef meets one of the two segments (ad) or (cd) .

Case (a): If fe meets (ad) in an event j (Figure 18) then, since there is no fastest path (Theorem 16), there is an event k and a path ek such that $[ajkd]$. The First Collinearity Theorem (Th.3) implies that ke meets af at an event l such that $[fla]$. Thus there is a path dl which, by part (i′) of this theorem, meets ae at an event m such that $[ame]$. The Third Collinearity Theorem applied to the kinematic triangle $\triangle ald$ with the path lk ($= ek$) such that $[akd]$ implies that $[aem]$. We have now obtained both $[ame]$ and $[aem]$ which contradict Theorem 1.

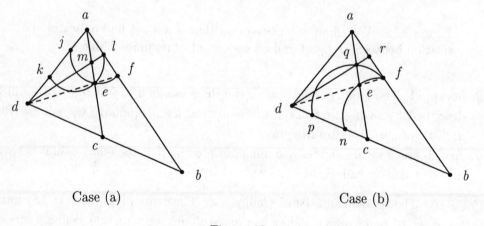

Case (a)   Case (b)

Figure 18

Case (b): If fe meets (cd) in an event n (Figure 18) such that $[dnc]$, then Theorem 16 implies the existence of an event p and a path fp such that $[dpn(c)]$,

40

and the Third Collinearity Theorem applied twice implies firstly the existence of an event q such that $[aqc]$ and $[pqf]$, and secondly that $[qec]$, whence $[aqec]$. Part (i') of this theorem implies the existence of a path dq which meets af $(= ab)$ in an event r such that $[arf]$. Now the three events f, q, p belong (respectively) to three distinct sides of the kinematic triangle $\triangle bdr$ and this contradicts Theorem 8. We have thus shown that both cases (a) and (b) lead to contradictions so there is no path from e to f.

To complete the proof of (ii) we will show that f is the closest event of $ab(e, \emptyset)$ to a by supposing the contrary, namely that there is an event $t \in ab(e, \emptyset)$ for which $[atf]$, and obtaining a contradiction. The definition of f implies that there is a path dt which, by part (i'), meets ac $(= ae)$ in an event u which, by the Third Collinearity Theorem, is between d and t. Now both d and t belong to $dt(e, \emptyset)$ but u does not, which contradicts Theorem 13. The proof of (ii) is now complete.

(ii') As in the proof of (ii) we show that there is no path from e' to f'. The contrary supposition that there is a path $e'f'$ implies that $e'f'$ meets one of the two segments (ad) or (cd), which are considered as cases (a'), (b') repectively.

Case (a'): The proof for this case is similar to the proof for Case (b) above. If $f'e'$ meets ad in an event n' (Figure 19) such that $[an'd]$ then Theorem 16 implies the existence of an event p' and a path $f'p'$ such that $[an'p'd]$. The Second Collinearity Theorem implies that ae' $(= ac)$ meets $f'p'$ in an event q' such that $[f'q'p']$ and $[ae'q']$ and the Third Collinearity Theorem implies that $[a(e')q'c]$. Part (i') of this theorem implies that dq' meets ab at an event r such that $[af'r'b]$. Thus $f'p'$ meets internal events of three sides of the kinematic triangle $\triangle adr'$, which contradicts Theorem 8.

Case (b'): The proof for this case is similar to the proof for Case (a) above. If $e'f'$ meets the segment (cd) in an event j' (Figure 19) such that $[dj'c]$ then Theorem 16 implies the existence of an event k' and a path $e'k'$ such that $[dk'j'c]$ whence the Second Collinearity Theorem implies that $k'e'$ meets $f'b$ in an event l such that $[f'l'b]$. Now part (i') of this theorem implies that dl' meets ac in an event m' such that $[ae'm'c]$, but the Third Collinearity Theorem implies that $[m'e'c]$ which is a contradiction of Theorem 1.

To complete the proof of part (ii') we use a proof which is similar to the last paragraph of the proof of part (i). We will show that f' is the closest event of $ab(e', \emptyset)$ to b by supposing the contrary, namely that there is an event $t' \in ab(e', \emptyset)$ for which $[bt'f']$, and obtaining a contradiction. The definition of f' implies that

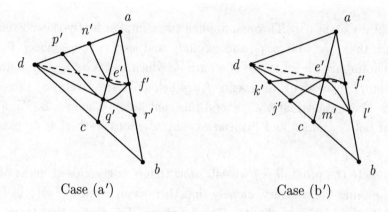

Case (a')　　　　　　　　　Case (b')

Figure 19

there is a path dt' which, by (i'), meets ac $(= ae')$ in an event u' which, by the Third Collinearity Theorem is between d and t'. Now both d and t' belong to $dt'(e', \emptyset)$ but u' does not, which contradicts Theorem 13. The proof of (ii') is now complete. *q.e.d.*

Theorem 20 (Corollary to First Optical Line Theorem)
Let abd' be a kinematic triangle, let c be an event such that $[bcd']$ and such that there is a path ac. Let d be an event of (ad'), let e be the closest event of $ac(d, \emptyset)$ to a, and let f be the closest event of $ab(d, \emptyset)$ to a.
If $[afb]$ then $[aec]$ and f is the closest event of $ab(e, \emptyset)$ to a (see Figure 20).

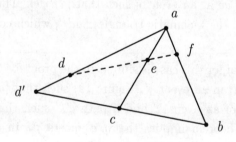

Figure 20

Proof (see Figure 21). By the Second Existence Theorem (Th.14) there is an event b' and a path db' such that $[afbb']$ and such that $ab(d, \emptyset) \subset (ab')$ and $ab(d', \emptyset) \subset (ab')$ so there are paths db' and $d'b'$. By the Second and Third Collinearity Theorems (Th.7 and Th.15), the path ac meets $(d'b')$ and so (ac) meets (db') in an event c'.

Now the configuration $adc'b'$ corresponds to the configuration $adcb$ of the previous theorem.

q.e.d.

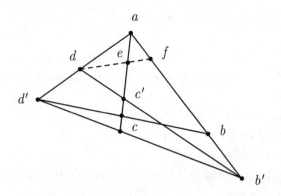

Figure 21

Theorem 21 *Each isotropy mapping is a bijection from $\mathcal{E}$ to $\mathcal{E}$.*

Proof In this proof we let the symbols θ, Θ, Q have the same meanings as in the statement of the Axiom of Isotropy (Axiom S). We must show that θ is both surjective and injective.

To demonstrate the surjective property, we observe that the Axiom of Connectedness (Axiom I2) together with the axioms of Existence and Dimension (Axioms I1 and I4) imply that, for any event a there is some path T such that $a \in T$. Since the induced map $\Theta : \mathcal{P} \to \mathcal{P}$ is bijective, there is some path T^* such that $\Theta : T^* \to T$. Now the mapping $\Theta : \mathcal{P} \to \mathcal{P}$ is induced by the mapping $\theta : \mathcal{E} \to \mathcal{E}$ so there must be an event $a^* \in T^*$ such that $\theta : a^* \mapsto a$.

Next we will show that θ is injective; that is we will show that any two distinct events a, b have distinct images $\theta(a)$, $\theta(b)$. The axioms of Existence, Connectedness and Dimension (Axioms I1,I2,I4) together with Theorem 16 imply the existence of at least two distinct paths through each event. The Axiom of Uniqueness (Axiom I3) implies that there is at most one path through two distinct events, so the sets of paths through the events a and b are distinct. Now if $\theta(a) = \theta(b)$, the sets of paths through $\theta(a)$ and $\theta(b)$ would not be distinct. This would contradict the Axiom of Isotropy (Axiom S(ii)) which states that the induced map $\Theta : \mathcal{P} \to \mathcal{P}$ is a bijection.

We have now shown that θ is both surjective and injective.

q.e.d.

43

Since both mappings $\theta : \mathcal{E} \to \mathcal{E}$ and $\Theta : \mathcal{P} \to \mathcal{P}$ are bijections and since they are so closely related, *we shall from now on use the lower case symbol to represent both mappings.* In subsequent theorems, other lower case Greek symbols such as ψ and ϕ will also be used for isotropy mappings.

Parallel Paths   In the next theorem we introduce the concept of parallelism, which will be discussed in much greater detail in Chapter 7. First we make the following definition:

Let abc be a kinematic triangle with $d \in (ab)$ and $t \in (ac)$. We will say that a path dt is *parallel through d to the ray* $\overrightarrow{bc}$ if :

(i) dt does not meet bc

(ii) for each u such that $[atuc]$, if there is a path du then du meets the ray $\overrightarrow{bc}$ at some event.

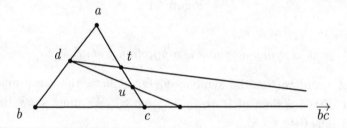

Figure 22   The path dt is parallel to the ray from b directed towards c

The next theorem (Th.22) establishes the existence of a parallel path under somewhat restrictive conditions.   (These conditions will be relaxed later in Theorem 56). Note that, as in absolute geometry, there is the possibility of two distinct parallels — one to the ray $\overrightarrow{bc}$ and the other to the ray $\overrightarrow{cb}$. It will be shown in the Euclidean Parallel Theorem (Th.68) that there is a unique parallel (as in Euclidean and affine geometry).

Theorem 22 (Existence of a Parallel Path)
Let abc be a kinematic triangle with an event $d \in (ab)$. If $ac(d, \phi)$ is not contained in the segment (ac), then there is an event $t \in (ac)$ and a path dt which is parallel to the ray $\overrightarrow{bc}$.

Proof Let p, q, r, s be events of $bc \setminus bc(d, \emptyset)$ such that $[srbcpq]$ (Figure 23). The First Collinearity Theorem (Th.3) implies that the paths dr, ds meet ac at events

r', s' such that $[ar's'c]$ and the Third Collinearity Theorem (Th.15) implies that the paths dp, dq meet ac at events p', q' which (by the same theorem) are ordered such that $[a(r')s'q'p']$. By Theorem 16 the two subsets of (ac), namely $\mathcal{A}$ which contains events satisfying the conditions of p' and q' and $\mathcal{B}$ (which contains events satisfying the conditions of r' and s') are disjoint and open.

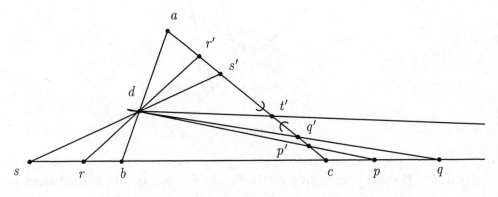

Figure 23

Now consider an unbounded sequence $(p_n : n = 1, 2, \ldots)$ of events of bc such that $p_1 = p$ and $p_2 = q$: then the corresponding sequence (p'_n) of events of (ac) (where $p'_1 = p'$ and $p'_2 = q'$) is contained in $\mathcal{A}$ and is bounded (by a subset which contains r' and s'). Thus by the Continuity Theorem (Th.12) there is a closest bound t' of $\mathcal{A}$ and, since $t' \notin ac(d, \emptyset)$,

(i) for all $p', q' \in \mathcal{A}$ and for all $r', s' \in \mathcal{B}$, $[ar's't'q'p'c]$
(ii) there is a path dt'.

Thus for any event u' such that $[at'u'c]$, there is a path du' which meets $\overrightarrow{bc}$ and (by (i)) the path dt' does not meet bc at any event. Thus dt' is parallel through d to the ray $\overrightarrow{bc}$.   q.e.d.

45

Theorem 23 *Let θ be an isotropy mapping as in the statement of the Axiom of Isotropy (Axiom S) and let R be a distinct path which meets Q at some event (not necessarily x). Then θ induces a strictly order-preserving bijection on (the set of events of) R.*

Proof Let $u = Q \cap R$ and take any events $v, w \in R$ such that $[uvw]$: we will first show that $[u, \theta(v), \theta(w)]$.

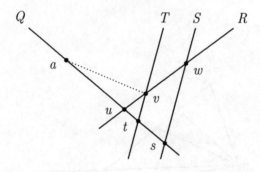

Figure 24

By Axiom I5 there is an event $a \in Q(v, \emptyset)$ (see Figure 24). The Second Existence Theorem (Th.14) implies that there is an event $s \in Q$ and a path $S := ws$ such that $[aus]$. Now the conditions of the preceding theorem (Th.22) are satisfied so there is a path T which passes through v, which is parallel to the ray $\overrightarrow{ws}$ of S and which meets Q at an event t such that $[auts]$.

Now the Axiom of Isotropy (Axiom S) implies that u, t, s are invariant and Theorem 21 implies that $u, \theta(v), \theta(w)$ are distinct and that the pair of paths $\theta(S), \theta(T)$ do not meet. Thus either (i) $[u, \theta(v), \theta(w)]$ or (ii) $[u, \theta(w), \theta(v)]$ or (iii) $[\theta(v), u, \theta(w)]$. For the orderings (ii) and (iii) the Collinearity Theorems (Th.7 and Th.15) would imply that $\theta(S)$ and $\theta(T)$ meet which would be a contradiction of the bijective property of θ (Theorem 21). It follows that $[u, \theta(v), \theta(w)]$.

The extension to the case $[vuw]$ is a consequence of the case above together with Theorem 10. $\hspace{3cm}$ *q.e.d.*

The next two theorems are called the "Reverse Record Function Theorem" and the "Record Function Theorem" (respectively). The concept of a "record function" is not defined or used in this book but the expression is incorporated in the names of the next two theorems, since it is related to a concept introduced by Milne (1935) and used previously by Walker (1948) and Schutz (1973, 1981). The related concepts of "modified record functions" are defined in Section 6.8.

Theorem 24 (Reverse Record Function)
Let Q, T be distinct paths which meet at an event x, let $c \in Q$ be an event distinct from x, let $b \in T$ be the closest event of $T(c, \emptyset)$ to x and let $a \in Q$ be the closest event of $Q(b, \emptyset)$ to x. Then $[xac]$.

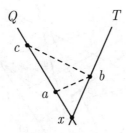

Figure 25

Remark   See Figure 25.

Proof   We will establish the result by supposing the contrary, namely that $a = c$, and obtaining a contradiction.

Let d be the furthest event of $Q(b, \emptyset)$ from x (Figure 26); then since there is no fastest path (Theorem 16) there is an event f and a path df such that $[xfb]$. By the Second Existence Theorem (Th.14) there is an event e and paths ce, de such that both $T(c, \emptyset)$ and $T(d, \emptyset)$ are between x and e. The Third Collinearity Theorem (Th.15) implies that the segments (ce) and (df) meet at an event g which is unreachable from b; since otherwise there would be a path joining b and g which would meet the segment (cd) whose end–events c and d both belong to $Q(b, \emptyset)$ and this would contradict Theorem 13. Let the closest event of $(fd)(b, \emptyset)$ to f be called h. Then $[dghf]$ and Theorem 13 implies the existence of a path $R := xh$ distinct from Q and T. By the Compact Collinear Set Theorem (Th.18) and the Third Collinearity Theorem it follows that R meets (ce) in an event i. Now the First Optical Line Theorem (Th.19(i),(ii)) applied to the configuration $xeic$ implies that there is an event $j \in (xi)$ such that j is the closest event of $R(c, \emptyset)$ to x and b is the closest event of $T(j, \emptyset)$ to x and $[xji]$. An argument similar to that of this paragraph (thus far) shows that there is an event j' such that j' is the closest event of $R(b, \emptyset)$ to x and c is the closest event of $Q(j', \emptyset)$ to x and the definition of R implies that $[xj'h]$. Now the definition of j implies that $[xjj']$ and the definition of j' implies that $[xj'j]$ whence $j = j'$ and $[xjh]$.

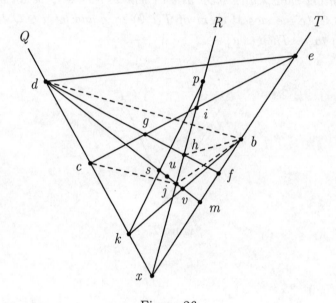

Figure 26

We next demonstrate that c is the closest event of $Q(c, R, x, \emptyset)$ to x; that is we show that there is no event k such that $[xkcd]$ for which there is an event $l \in R(c, \emptyset)$ such that $k \in Q(l, \emptyset)$. By the First Optical Line Theorem (Th.19(i)) any such event l would be between j and i, and Theorem 4 and the Prolongation Theorem (Th.6) imply the existence of an event p and a path kp such that $[xjlip]$. By the First Optical Line Theorem and the Compact Collinear Set Theorem there is a path dj which meets T in an event m such that $[xmf]$ and which, by the Third Collinearity Theorem meets (kp) in an event s. The First Optical Line Theorem applied to the configuration $jpsk$ would then imply the existence of an event $u \in (sj)$ which is unreachable from k. Now $(sj) \subset (dj)$ so (sj) is unreachable from b and, by supposition, there is a path bk which, by the Third Collinearity Theorem, meets $dm(= dj = um)$ in an event v such that $[djvm]$ and $[bvk]$: but then the unreachable set $bk(u, \emptyset)$ would be disconnected (by v) which would contradict Theorem 13. Thus $Q(c, R, x, \emptyset) = \{c\}$.

We have not shown that $Q(c, T, x, \emptyset) = \{c\}$, so we can not apply the Axiom of Isotropy (Axiom S) to the paths Q, R, T. However if we apply the same argument as in the preceding paragraphs to the paths Q, R and the events c and j (rather than to the paths Q, T and the events c and b) and if we denote the corresponding symbols with asterisks (where now $T^* := R$ and $b^* := j$) then we observe that both $Q(c, R^*, x, \emptyset)$ and $Q(c, T^*, x, \emptyset)$ are equal to the singleton set $\{c\}$. Furthermore there is a path d^*j^* but there is no path d^*b^*.

The Axiom of Isotropy (Axiom S) can now be applied to the paths Q, R^*, T^* with the event c taking the place of Q_a, with R^* taking the place of R and with T^* taking the place of S in the statement of the axiom. By Theorem 23 there is an isotropy mapping θ which is order-preserving and bijective from the events of R^* to the events of T^*: thus θ maps j^* onto b^*. However there is a path d^*j^* but there is no path from d^* to b^*: this contradicts the bijective property of the induced mapping from $\mathcal{P}$ to $\mathcal{P}$ (Axiom S(ii)). This contradiction completes the proof. $\qquad$ *q.e.d.*

Theorem 25 (Record Function)
Let Q, T be distinct paths which meet at an event x, let $a \in Q$ be an event distinct from x, let $b' \in T$ be the furthest event of $T(a, \emptyset)$ from x and let $c \in Q$ be the furthest event of $Q(b', \emptyset)$ from x.
Then $[xac]$.

Remark   See Figure 27a.

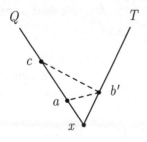

Figure 27a

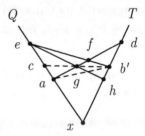

Figure 27b

Proof By Theorem 4 unreachable sets are bounded, so there is an event d and a path ad such that $[xb'd]$ (Figure 27b) and there is an event e and a path $b'e$ such that $[xae]$. The Third Collinearity Theorem (Th.15) implies that (ad) meets $(b'e)$

in an event f. The Reverse Record Function Theorem (Th.24) implies that the closest event of $ad(b', \emptyset)$ to d is an event g such that $[agd]$ and by Theorem 13, $[agfd]$. Also by Theorem 13, there is a path eg which, by the Second Collinearity Theorem (Th.7), meets xd in an event h such that $[xhb'd]$. Now the First Optical Line Theorem (Th.19) implies the existence of an event $c \in (xe)$ such that c is the closest event of both $xe(b', \emptyset)$ and $xe(g, \emptyset)$ to e, whence $[eca]$. But there is a path $ag(= ad)$ so $c \neq a$ and so $[eca]$, which implies $[xac]$. q.e.d.

4.6 Causality theorem

The Causality Theorem will enable us to deduce important properties of a partial order relation, which will be called "temporal order", in the Collinear Set Theorem (Th.36) of the following Chapter 5. The Causality Theorem (Th.26) may be described in an intuitive manner by stating that "optical lines do not cross within a kinematic triangle".

Theorem 26 (Causality)

(i) *For each kinematic triangle, there is exactly one vertex which has its unreachable set contained within the opposite side.*

(ii) *For each kinematic triangle, there are exactly two vertices such that each internal event may be (separately) joined to each of the two vertices by a single path. Thus no internal event is unreachable from two vertices.*

Proof The proof of (i) consists of two steps: in Step 1 we will show that there is at most one vertex with its unreachable set on the opposite side and in Step 2 we will show that there is one such vertex.

Step 1: Let abc be a kinematic triangle with $ac(b, \emptyset)$ contained in the segment (ac). We will show that no event of $bc(a, \emptyset)$ belongs to the segment (bc) (and by symmetry it will then follow that no event of $ab(c, \emptyset)$ belongs to the segment (ab)).

Let d be the closest event of $ac(b, \emptyset)$ to a and let e be the closest event of $ab(d, \emptyset)$ to a (Figure 28): the Reverse Record Function Theorem (Th.24) implies that $[aeb]$. Since there is no fastest path (Theorem 16) there is an event f between b and c such that $bc(a, \emptyset)$ is not contained in the segment (fc). By the Optical Line Theorem (Th.19(ii)) there is an event $h \in (af)$ such that h is the closest event of $af(b, \emptyset)$ to a and such that d is the closest event of $ac(h, \emptyset)$ to a and b, h, d are in optical line. By Theorem 20 there is an event i which is the closest event of $af(d, \emptyset)$ to a and

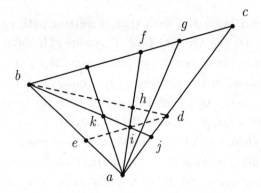

Figure 28

such that e is the closest event of $ab(i,\emptyset)$ to a and d,i,e are in optical line. Now the Reverse Record Function Theorem (Th.24) implies that $[aih]$ and so there is a path bi which (by the Optical Line Theorem (Th.19(i'))) meets ac in an event j such that $[ajdc]$. Theorem 13 implies that for each event $k \in (bi)$ there is no path from k to d and hence that there is a path ak. Now the Optical Line Theorem (Th.19(i)) implies that $bc(a,\emptyset)$ is not contained in the segment (bf). We have now shown that $bc(a,\emptyset)$ is not contained in the segment (bc), which completes the proof of Step 1.

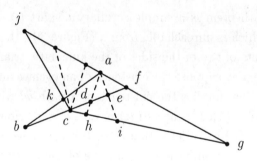

Figure 29

Step 2: Let abc be a kinematic triangle with $bc(a,\emptyset)$ not contained in the segment (bc): we will show that one of the other two vertices has its corresponding unreachable set located (internally) on the opposite side. Without loss of generality we suppose that $bc(a,\emptyset)$ lies on the prolongation of (bc) beyond c and we let d,e be (respectively) the closest and furthest events of $bc(a,\emptyset)$ to (resp. from) c (Figure 29). By Theorem 4 there is an event f and a path af such that $[bcdef]$. Now let g be an arbitrary event on af such that $[afg]$ and such that there is a path cg (in accordance with Theorem 14): then by the Optical Line Theorem (Th.19(i)) the unreachable set $cg(a, \emptyset)$ is contained in the segment (cg) and hence Step 1 of this proof implies that $af(c, \emptyset)$ is not contained in the segment (ag). But g is arbitrary, so $af(c, \emptyset)$ is not contained in the ray $\overrightarrow{af}$. Again by Theorem 4 there is an event j and a path cj such that $[jaf]$ with $af(c, \emptyset)$ $(= aj(c, \emptyset))$ contained in the segment (aj). By the Third Collinearity Theorem (Th.15), the segment (cj) meets (ab) at an event k and then the Optical Line Theorem (part (i)) implies that $ab(c, \emptyset)$ is contained in the segment (ak) which is contained in the segment (ab). This completes Step 2 of the proof of (i).

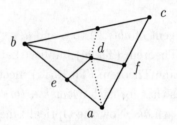

Figure 30

Part (ii) of the theorem is a simple corollary of part (i). Let d be an internal event of $\mathcal{C}(a, b, c)$ which is unreachable from a (Figure 30); then d belongs to a path which connects events of two of the sides of the kinematic triangle $\triangle abc$, so without loss of generality we may suppose the existence of an event e and a path de such that $[aeb]$. Theorem 13 then implies the existence of a path bd which, by the Compact Collinear Set Theorem (Th.18) meets (ac) in an event f. Now by the First Optical Line Theorem (Th.19(i)) we observe that $bc(a, \emptyset)$ is contained in the segment (bc). If any event of $\mathcal{C}(a, b, c)$ was unreachable from one of the other two vertices, a similar argument would show that the second vertex would also have its unreachable set within the opposite side, which would contradict part (i) of this theorem. The proof of (ii) is now complete. *q.e.d.*

5. Existence and properties of collinear sets

The purpose of this chapter is to describe "collinear sets" which may be thought of intuitively as sets of paths in "one-dimensional motion" together with the sets of events which belong to them. In Section 5.1 we establish some relationships between unreachable sets: these relationships are developed further in the discussion of the order properties of "signal functions" in the next chapter. In Section 5.2 we define the partial order relation of "temporal order" on a compact collinear set. The theorems of the next two sections will lead to Section 5.5 and in particular to the Collinear Set Theorem (Th.36) which describes the existence, uniqueness and temporal order properties of collinear sets.

5.1 Order properties of unreachable sets

The next two theorems describe properties of order relating the unreachable sets of two paths. In Section 6.2 of the next chapter we will define the concept of "signal functions" and then Theorem 28 of the present section will become superseded by Theorem 38 which states that "signal functions are weakly order-preserving".

Theorem 27 *Let Q, R meet at x, let $y \in Q$ be an event distinct from x, and let a, b be events of R.*
If $[x, a, R(y, \emptyset), b]$, then $[x, Q(a, \emptyset), y, Q(b, \emptyset)]$ (see Figure 31).

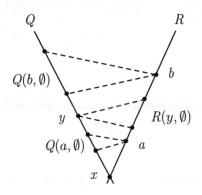

Figure 31

Proof We first show that $[x, y, Q(b, \emptyset)]$ by supposing the contrary, namely that $[x, Q(b, \emptyset), y]$ or that $[y, x, Q(b, \emptyset)]$, and then obtaining a contradiction. Now there is a path yb and Theorem 14 implies the existence of an event c such that $[cxy]$ and such that $[c, Q(b, \emptyset), y]$. But the kinematic triangle $\triangle ybc$ with the unreachable sets corresponding to y and b contradicts the Causality Theorem (Th.26(ii)), so we conclude that $[x, y, Q(b, \emptyset)]$.

Next we show that $[x, Q(a, \emptyset), y]$ by supposing the contrary and obtaining a contradiction. The contrary supposition includes two cases (i) $[x, y, Q(a, \emptyset)]$ and (ii) $[Q(a, \emptyset), x, y]$ which must be considered separately.

In Case (i), Theorem 4 implies the existence of an event c such that $[x, y, Q(a, \emptyset), c]$ (Figure 32(i)) and the Second Existence Theorem (Th.14(iii)) implies the existence of an event d and paths yd, cd such that $[x, a, R(y, \emptyset), d]$. Therefore there is a path ac which, by the Third Collinearity Theorem (Th.15) meets (yd) in an event e. Now the Optical Line Theorem (Th.19(i)) implies that $yd(a, \emptyset)$ is contained in the segment (ye) which is contained in the segment (yd). This contradicts the Causality Theorem (Th.26) for the kinematic triangle $\triangle yad$.

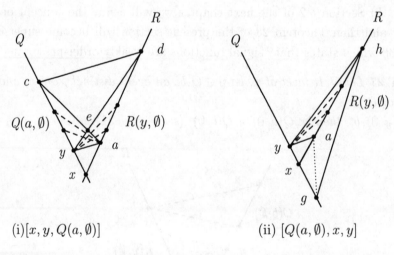

(i) $[x, y, Q(a, \emptyset)]$　　　　(ii) $[Q(a, \emptyset), x, y]$

Figure 32

In Case (ii) we consider any event $g \in Q(a, \emptyset)$ and then the Second Existence Theorem (Th.14(iii)) implies the existence of an event $h \in R$ and paths gh, yh such that $[x, a, R(y, \emptyset), h]$. But this configuration (Figure 32(ii)) contradicts the Causality Theorem (Th.26(ii)). The proof is now complete.　　　　*q.e.d.*

The next theorem describes order relationships between unreachable sets. It will be superseded in Section 6.2 of the next chapter by Theorem 38 which states that "signal functions are weakly order-preserving".

Theorem 28 *Let Q, R meet at x and let w, y, z be events of Q with w' (resp. w'') being the closest (resp. furthest) event of $R(w, \emptyset)$ from (resp. to) x and with y', y'', z', z'' being similarly defined. Then*
   (i) $[xyz] \Longrightarrow [xy''z'']$
   (ii) $[xyz] \Longrightarrow [xy'z']$
   (iii) $[wxy] \Longrightarrow [w''w'xy'y'']$.

Remark   See Figure 33.

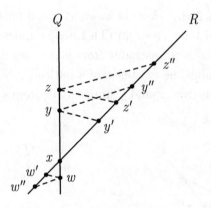

Figure 33

Proof   (i) We suppose the contrary; that is we suppose that either (a) $[x \ z'' \ y'']$ or (b) $[y'' \ x \ z'']$ and in both cases we will obtain contradictions. (See Figure 34).

In Case (a) there is a path zy'' and the configuration contradicts the Causality Theorem (Th.26(i)). In Case (b), Theorem 13 implies that $[y''y'xz'z'']$ so there is a path zy' and by the Prolongation Theorem (Th.6) there is an event a and a path za such that $[xz''a]$: but this configuration contradicts the Causality Theorem (Th.26(ii)).

(ii) We suppose the contrary; that is, we suppose that $[xz'y']$ or $[y'xz']$ and will then obtain a contradiction. The result (i) above, together with Theorem 13 excludes the case $[y'xz']$ so we proceed to consider the case $[xz'y']$ (Figure 35). By the Second

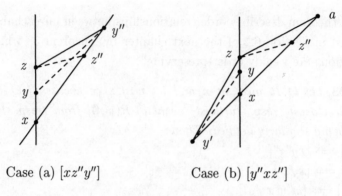

Case (a) $[xz''y'']$ Case (b) $[y''xz'']$

Figure 34

Existence Theorem (Th.14(ii)) there is an event a and paths ya, za such that $[xz''a]$ and $[xy''a]$. The Optical Line Theorem (Th.19(ii)) implies the existence of an event $b \in (ya)(z, \emptyset)$ such that z' is unreachable from b and therefore b is unreachable from z'. The supposition implies the existence of a path yz'. Now the kinematic triangle $\triangle ayz'$ with b unreachable from z' and y' unreachable from y contradicts the Causality Theorem.

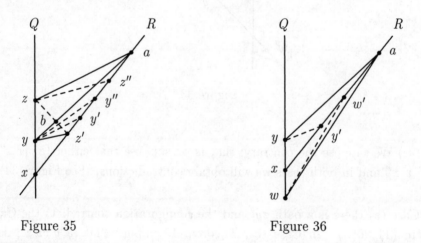

Figure 35   Figure 36

(iii) Since unreachable sets are connected (Theorem 13) it is sufficient to show that $[w'xy']$: we will do this by supposing the contrary, namely that $[xy'w']$ or $[xw'y']$, and obtaining a contradiction.

By the Second Existence Theorem there is an event a and paths wa, ya such that $[xy'a]$ and $[xw'a]$ (Figure 36). The kinematic triangle $\triangle ayw$ with the internal events

w' unreachable from w and y' unreachable from y contradicts the Causality Theorem (part (ii)): this completes the proof of (iii). *q.e.d.*

Theorem 10 implies that the events on any path are linearly ordered so the three–term betweenness relation $[\cdots]$ can be replaced by an irreflexive binary linear order relation $<$ and similarly the non-strict betweeness relation $[\![\cdots]\!]$ can be replaced by a reflexive binary linear order relation $\leq$. If $a < b$ we say that a is *before* b, that a *precedes* b, or that b is *after* a. We define the symbols $>$ and $\geq$ so that the statement " $b > a$ " is equivalent to " $a < b$ " and so that " $d \geq c$ " is equivalent to " $c \leq d$ ". The relations $<$ and $\leq$ will be called *temporal order relations*. The definitions can be extended to apply to sets of events (as in the definition preceding the Second Existence Theorem (Th.14)).

5.2 Partial order on a compact collinear set

It will be possible to define a linear order relation $<$ on each path which meets two distinct events of a compact collinear set $\mathcal{C}(a, b, c)$ in such a way that the relation becomes a partial order relation on all events of $\mathcal{C}(a, b, c)$: this result will be established in the next theorem. Before proving this theorem, each path (which has an interval belonging to $\mathcal{C}(a, b, c)$) must be assigned a "sense of direction" for its linear order relation, so we make the definitions which follow. For convenience and without loss of generality, we consider a kinematic triangle $\triangle abc$ such that $ac(b, \emptyset)$ is between a and c (Figure 37). Then

(i) for any path R which passes through a and which meets the side $|bc|$ we assign a linear order on R such that $a < R \cap |bc|$.

(ii) for events $f, g \in \mathcal{C}(a, b, c) \setminus \{a\}$ the Causality Theorem (Th.26(ii)) implies that there are paths af and ag. If

$$ag(f, \emptyset) \subset (ag)$$

there is a path fg and the Compact Collinear Set Theorem (Th.18) implies that fg meets $\mathcal{C}(a, b, c)$ in an interval: on the path fg we assign an order such that $f < g$.

In the following theorem we show that the linear order relations on the set of paths (which meet $\mathcal{C}(a, b, c)$ in intervals) is a partial order relation when restricted to the events of $\mathcal{C}(a, b, c)$.

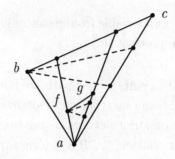

Figure 37

Theorem 29 (Partial Order)

(i) *The relation $<$ is a partial order relation on $\mathcal{C}(a, b, c)$.*

(ii) *For events $f, g \in \mathcal{C}(a, b, c)$ such that $f < g$, there is a path fg which meets the boundary in two events and all events of the corresponding interval belong to $\mathcal{C}(a, b, c)$.*

Remark After the proof of this theorem we will be able to use the symbol $<$ in the usual manner for a partial order relation (on a compact collinear set): thus for $f, g, h \in \mathcal{C}(a, b, c)$ such that $f < g$ and $g < h$ we will write

$$f < g < h$$

where the transitivity is suggested by the notation. (Within the proof of this theorem the same convention is used but only for events which belong to a single path, where the relation $<$ denotes a linear ordering as established by Theorem 10).

Proof (i) As in the preceding definition we suppose (without loss of generality) that $ac(b, \emptyset)$ is between a and c. The definition of $<$ is irreflexive and the Causality Theorem (Th.26(i)) (together with the definition (part(ii)) preceding the present theorem) implies that the relation is asymmetric. To establish that the relation $<$ is transitive we consider any three events f, g, h of $\mathcal{C}(a, b, c)$ such that af, ag, ah meet bc at events i, j, k respectively and such that $f < g$ and $g < h$. We must show that $f < h$ and so we consider the following cases.

Case (a) $i < j < k$ (Figure 38). The Causality Theorem (Th.26(ii)) implies the existence of a path fc, which (by the Third Collinearity Theorem (Th.15)) meets (ak) in an event d. Furthermore the Causality Theorem (part (ii)) implies both that $af(c, \emptyset)$ is not contained in the segment (af) and that $fc(a, \emptyset)$ is not contained in the segment (fc). Let l be the furthest event of $ag(f, \emptyset)$ from a and let m be the furthest event of $ak(f, \emptyset)$ from a: then the Causality Theorem (part(i)) implies that $ak(f, \emptyset)$ is contained in the segment (ad) and the Optical Line Theorem (Th.19(ii$'$)) implies that m is the furthest event of $ak(l, \emptyset)$ from a.

Let n be the furthest event of $ak(g, \emptyset)$ from a. Since $f < g$ and $g < h$ it follows that $l < g$ and $n < h$. Since signal functions are weakly order-preserving (Theorem 28), the relation $l < g$ implies $m \leq n$. Now by Theorem 10, $m \leq n$ and $n < h$ imply $m < h$, so there is a path fh and $f < h$.

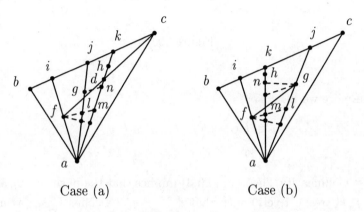

Case (a)   Case (b)

Figure 38

Case (b) $i < k < j$ (Figure 38). Let l, m, n be defined as in Case (a): then l is the furthest event of $aj(m, \emptyset)$ from a. Theorem 27 implies that $m < n$. On the path ak, Theorem 10 implies that $a < m < n < h$ (whence $m < h$) so there is a path fh and $f < h$.

Case (c) $j < i < k$ (Figure 39). Let u be the furthest event of $ak(f, \emptyset)$ from a, let v be the furthest event of $ai(g, \emptyset)$ from a and let w be the furthest event of $ak(g, \emptyset)$ from a. Then as in Case (a), w is the furthest event of $(ak)(v, \emptyset)$ from a. Now $f < g$, so Theorem 27 implies that $f < v$. Since signal functions are weakly order-preserving, this implies that $u \leq w$. Also $g < h$ implies that $w < h$ so it follows that $u < h$, whence there is a path fh and $f < h$.

Case (d) $k < j < i$ (Figure 39). Let us denote c as b': then the Second Existence Theorem (Th.14) implies the existence of an event c' such that $[a, ab(b', \emptyset), c']$. If we now assign an order relation $<'$ on $\mathcal{C}(a, b', c')$ in a manner analogous to that for the relation $<$ on $\mathcal{C}(a, b, c)$ then the restriction of $<'$ to $\mathcal{C}(a, b, c)$ is the same relation

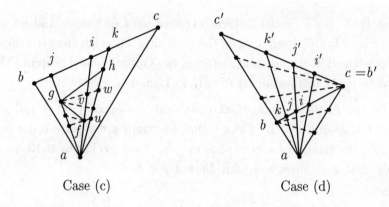

Case (c)   Case (d)

Figure 39

as $<$: furthermore we have

$$a < b, \qquad b < c, \qquad a < c$$
$$a <' b', \qquad b' <' c', \qquad a <' c' .$$

Now the First Collinearity Theorem (Th.3) implies that the paths ai, aj, ak meet $b'c'$ in events i', j', k' (respectively) such that $i' <' j' <' k'$. Thus Case (a) now applies for f, g, h regarded as events of $\mathcal{C}(a, b', c')$, so $f <' h$ and there is a path fh. But $f, g, h \in \mathcal{C}(a, b, c)$ whence $f < h$, since the restriction of $<'$ to $\mathcal{C}(a, b, c)$ is $<$.

Case (e) $j < k < i$.

Case (f) $k < i < j$.

For both of these cases an extended compact collinear set $\mathcal{C}(a, b, c)$ is defined as in Case (d). Then as in the proof of Case (d), we can regard Case (e) (and Case (f), respectively) as Case (b) (and Case (c), respectively) in the extended set $\mathcal{C}(a, b', c')$.

Cases remaining: the remaining cases where $i = j$ or $j = k$ can be considered as special sub–cases of Case (a) and Case (d).

(ii) This is an immediate consequence of (i) and the Compact Collinear Set Theorem (Th.18).   *q.e.d.*

5.3 Order properties which imply coincidence

In the proof of the preceding theorem (Th.29) we saw that a partial order relation $<$ on a compact collinear set has a unique extension to a larger compact collinear set. From now on we shall use the same symbol $<$ for both sets (instead of $<$ for the smaller set and $<'$ for the larger set).

On any path which meets a compact collinear set in an interval, there is a linear order relation $<$ on the path which conforms with the partial order relation $<$ on the interval. In subsequent theorems we will use these linear orderings on the sets of events of paths but we will not use them between paths; that is, we do not assume that they combine to form a partial ordering. This important result will be proved in the Collinear Set Theorem (Th.36).

Theorem 30 (Ordered Coincidence)
Let Q, R be distinct paths which meet at x. Let $a, e \in Q$ and let $b, d \in R$ with xde being a kinematic triangle with an ordering such that $x < d < e$.

(i) *If $x < a < b < d < e$, then the paths ab and ed meet at an event c. Furthermore there is a compact collinear set in which.*

$$x < a < b < c < d < e.$$

(ii) *If $a < x$ and $b < x$ and $[b, R(a, \emptyset), x]$ then the paths ab and ed meet at an event c. Futhermore there is a compact collinear set in which $b < a < c < d < e$.*

Remark   See Figures 40(i) and 40(ii).

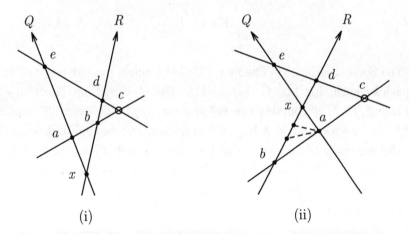

Figure 40

Proof (i) (See Figure 41). Theorem 27 applied to the paths ex and ed implies that $[e, d, ed(x, \emptyset)]$ and, since unreachable sets are bounded (Theorem 4), there is an event f and a path xf such that $[e, d, ed(x, \emptyset), f]$. Now the compact collinear set $C(f, x, e)$ contains $C(x, d, e)$ and since $x < e$ in $C(x, d, e)$, it follows that in the larger set $f < x < e$ whence

$$f < x < a < b < d < e.$$

The Compact Collinear Set Theorem (Th.18) and the Third Collinearity Theorem (Th.15) imply that ab meets the interval (fd) at an event c such that $a < b < c$ (and $c < d < e$), which completes the proof of (i).

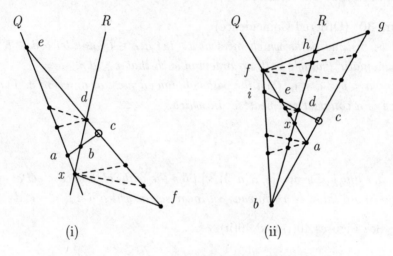

Figure 41

(ii) The Second Existence Theorem (Th.14) implies the existence of an event f and a path bf such that $[axef]$ (Figure 41). The Optical Line Theorem (Th.19(i)) implies that $bf(a, \emptyset)$ is contained in the segment (bf). Theorem 27 applied to the paths bf, ba implies that $[b, a, ba(f, \emptyset)]$ and then the Second Existence Theorem implies the existence of an event g and a path fg such that $[b, a, ba(f, \emptyset), g]$. The Second Collinearity Theorem (Th.7) implies that bx meets (fg) in an event h. Thus events in the compact collinear set $\mathcal{C}(b, f, g)$ are ordered so that $b < f < g$ and since $<$ is a partial order on this set we have

$$b < a < x < d < e < f < h < g \ .$$

In the subset $\mathcal{C}(a, f, g)$, the event e is on the boundary while the event d is internal and precedes e so the ray $\overrightarrow{ed}$ meets the boundary at an event c before d; that is

$$a < c < d < e < g$$

which, together with the previous partial order relations, completes the proof of (ii).

q.e.d.

5.4 Collinear paths which meet at a single event

Given three distinct paths Q, R, S which meet at an event x, we say that the *path R is between the paths Q and S* and we write $\langle Q, R, S \rangle$ if:

(i) for all $a \in Q$, $c \in S$ such that $x < a < c$, the interval $R \cap C(x, a, c)$ has a corresponding segment of internal events,

(ii) for all $a' \in Q$, $c' \in S$ such that $x > a' > c'$, the interval $R \cap C(x, a', c')$ has a corresponding segment of internal events.

Part (i) applies to events after the event x of coincidence while (ii) applies to events before the event x of coincidence.

Theorem 31 (Intermediate Path)
Let Q and S be two distinct paths which meet at some event x, and let T be a path which meets Q and S at events a and c, respectively, after x. For each event $b \in T$ such that $[abc]$, there is a path R which passes through x and b such that

$$\langle Q, R, S \rangle .$$

Remark The word "after" can be replaced by the word "before" in the statement of the theorem.

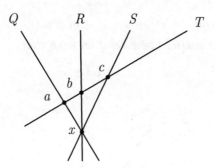

Figure 42   The Intermediate Path Theorem

Proof (i) Without loss of generality we suppose that $x < a < c$. Then for the given compact collinear set $C(x, a, c)$, the events of the segment (xb) are internal by definition. For a larger compact collinear set $C(x, a'', c'')$, with $x < a < a''$ and $x < c < c''$ the event b is internal and the Compact Collinear Set Theorem implies that property (i) of the definition is satisfied.

(ii) It is sufficient to show that there are events $j \in Q, k \in R, l \in S$ (analogous to a, b, c respectively) such that $[jxa]$, $[kxb]$, $[lxc]$ and $[jkl]$ (as in Figure 43).

63

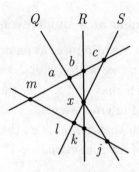

Figure 43

The Prolongation Theorem (Th.6) implies the existence of an event $m \in ac$ such that $[mabc]$. The Existence Theorem (Th.14) implies the existence of an event $j \in Q$ and a path mj such that $[jxa]$. The Second Collinearity Theorem (Th.7) now implies the existence of events $k \in R$, $l \in S$ such that $[mlkj]$, $[cxl]$ and $[bxk]$.   q.e.d.

Theorem 32 (Coincidence Corollary)
Let Q, R, S be distinct paths (which meet at x) such that $\langle Q, R, S \rangle$. Let T be a path which meets Q at a and R at b.
If $x < b < a$ then T meets S at an event c and there is a compact collinear set $\mathcal{C}(x, c, a)$ with an induced partial order relation such that

$$x < c < b < a .$$

Remark   See Figure 44.

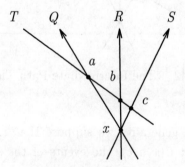

Figure 44

Proof   By the Second Existence Theorem (Th.14) there is an event $d \in S$ such that $[x, S(a, \emptyset), d]$ whence, in $\mathcal{C}(x, a, d)$, we have $x < a < d$. Also, the betweenness relation

64

$\langle Q, R, S \rangle$ implies that the path R meets the segment (ad) in an event e such that $(x <)a < e < d$. Now Theorem 27, together with $x < b < a$ and $x < a < e$, implies $x < b < a < e$ whence $[xbe]$ and so $b \in \mathcal{C}(x, a, d)$. Finally the Second Collinearity Theorem (Th.7), applied in $\mathcal{C}(x, a, d)$ to the triangle $\triangle xde$ with $[dea]$ and $[ebx]$, implies that ab meets $xd(= S)$ in an event c such that $[xcd]$ and $[abc]$.   q.e.d.

If Q, S are distinct paths which meet at an event x, we define the *collinear sub–SPRAY*

$$CSP\langle Q, S \rangle := \{R : \langle R, Q, S \rangle, \langle Q, R, S \rangle, \langle Q, S, R \rangle; R \in SPR[x]\} \cup \{Q, S\} .$$

Theorem 33 (Each CSP is simply ordered)
Let Q, S be distinct paths which meet at x. Then
  (i) $CSP\langle Q, S \rangle$ *is a simply ordered set*
  (ii) *for any distinct paths $U, V \in CSP\langle Q, S \rangle$,*

$$CSP\langle U, V \rangle = CSP\langle Q, S \rangle ,$$

  (iii) $CSP\langle Q, S \rangle$ *has no bounding paths.*

Proof (a) By the Intermediate Path Theorem (Th.31) the set of paths between Q and S is simply ordered.

(b) Let $U, V \in CSP\langle Q, S \rangle$ be such that $\langle Q, S, U \rangle$ and $\langle Q, S, V \rangle$ (Figure 45). Let W be a path which meets Q, S at events a, b respectively such that $a > b > x$. By the preceding theorem (Th.33), the path W meets U, V at events c, d (respectively) such that (using the order relations $<$ and $\leq$ on the events of the path W)

$$a > b > c \quad \text{and} \quad a > b > d .$$

Since W is a linearly ordered set, either

$$a > b > c > d \quad \text{or} \quad a > b > d \geq c$$

so by the Intermediate Path Theorem either

$$\langle Q, S, U, V \rangle \quad \text{or} \quad \langle Q, S, V, U \rangle \quad \text{or} \quad U = V .$$

(c) Consider $P \in CSP\langle Q, S \rangle$ such that $\langle P, Q, S \rangle$ and let V be as in Part (b) above. Let T be a path which meets U, S, Q in $\mathcal{C}(x, c, a)$ at e, f, g (respectively) such that

$$c > e > f > g > x . \tag{1}$$

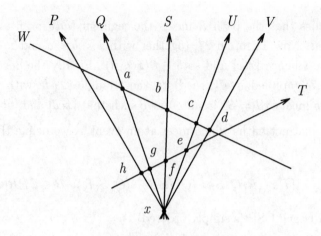

Figure 45

By the preceding theorem, T meets P at an event h such that

$$x < h < g < f \tag{2}$$

in $\mathcal{C}(x, h, f)$, so by (1) and (2) the events of the path T are ordered such that

$$h < g < f < e.$$

Finally, by the Intermediate Path Theorem, this implies that $\langle P, Q, S \rangle$, $\langle P, Q, U \rangle$, $\langle P, S, U \rangle$ and $\langle Q, S, U \rangle$. This completes the proof of (i). Part (ii) is an immediate corollary of (i). Part (iii) is a consequence of (ii) and Theorem 16 applied to the paths W and T. *q.e.d.*

As a consequence of the preceding theorem we make the following definition: given any four paths Q, S, U, V which meet at some event x, we write

$$\langle Q, S, U, V \rangle \iff \langle Q, S, U \rangle \text{ and } \langle Q, S, V \rangle \text{ and } \langle Q, U, V \rangle \text{ and } \langle S, U, V \rangle.$$

The notation $\langle \cdots \rangle$ may be extended in the obvious way to any simply ordered set of paths, provided that every ordered triple satisfies the definition of betweenness: in particular the definition applies for sets of paths with linearly ordered indexing sets.

The next two theorems (Th.34 and 35) are established for the sole purpose of proving the important Collinear Set Theorem (Th.36) of the final section of this chapter. Theorem 34 is illustrated in Figure 46 and is called the "Forward Collinear sub–spray" theorem.

Theorem 34 (Forward Collinear sub–spray)

Let Q, S be two distinct paths which meet at an event x. Let

$$csp \langle Q, S \rangle := \{R_y : R_y \in R, R \in CSP \langle Q, S \rangle\} .$$

Then there is a subset $csp^+ \langle Q, S \rangle$ of $csp \langle Q, S \rangle$ and a partial order relation $<$ on $csp^+ \langle Q, S \rangle$ such that:

(i) *for all $R_y \in csp^+ \langle Q, S \rangle \setminus \{x\}$, $x < R_y$*
(ii) *for all $a, b \in csp^+ \langle Q, S \rangle$*

$$a < b \implies \text{there is a path } ab$$

(iii) *for any path T which meets $csp^+ \langle Q, S \rangle$ in two distinct events, there is an induced linear ordering $<$ on T and the subset*

$$\{T_z : T_z > x, \ T_z \in T\}$$

is contained in $csp^+ \langle Q, S \rangle$.

Remark   The event x belongs to $csp^+ \langle Q, S \rangle$.

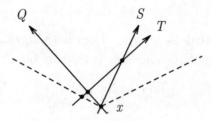

Figure 46

Proof (i) (a) The concept of a compact collinear set (developed in Theorem 18) can be extended to non-compact sets in the following way. Let Q, S be distinct paths which meet at an event x and let $(Q_m : m = 1, 2, 3, \cdots)$ and $(S_n : n = 1, 2, 3, \cdots)$ be unbounded monotonic sequences of events after x. Then for each Q_m there is some S_n such that $[x, S(Q_m, \emptyset), S_n]$ so for all $m' > m$ and for all $n' > n$ such that $[x, S(Q_{m'}, \emptyset), S_{n'}]$,

$$C(x, Q_m, S_n) \subset C(x, Q_{m'}, S_{n'})$$

and the partial order relation $<$ of Theorem 29 extends in a unique manner from the smaller set to the larger set. Next we define

$$C^+(Q, S) := \bigcup C(x, Q_m, S_n)$$

67

where the union is taken over all m, n such that $[x, S(Q_m, \emptyset), S_n]$. (Note that $C^+(Q, S) = C^+(S, Q)$ as a consequence of Theorems 27 and 4.) Thus $C^+(Q, S)$ is partially ordered by the induced relation $<$ and has the property that, for any events $a, b \in C^+(Q, S)$ such that $a < b$,

$$|ab| \subseteq C^+(Q, S) \,.$$

(b) We next show that for any path R such that $\langle Q, R, S \rangle$,

$$C^+(Q, R) \subset C^+(Q, S) \,.$$

Define an unbounded monotonic sequence of events $(R_l : l = 1, 2, 3, \ldots)$: then

$$C^+(Q, R) = C^+(R, Q) = \bigcup C^+(x, R_l, Q_m)$$

(where the union is specified in much the same way as before). Now for any $C(x, R_l, Q_m)$, Theorem 14 implies that we can take $Q_{m'}$ such that
(i) $Q_{m'} > Q_m$ and
(ii) $Q_{m'} > R_l$

and then take S_n such that $S_n > Q_{m'}$. Then both $R_l, Q_m \in C(x, Q_{m'}, S_n)$ whence $C(x, R_l, Q_m) \subset C(x, Q_{m'}, S_n)$. Since this is true for any $C(x, R_l, Q_m)$ it follows that

$$C^+(Q, R) \subset C^+(Q, S) \,.$$

(c) Hence for any paths U, V such that $\langle U, Q, S, V \rangle$ it follows that

$$C^+(Q, S) \subset C^+(U, V)$$

and the partial order relation $<$ extends uniquely from the smaller set to the larger set, so we define

$$csp^+ \langle Q, S \rangle := \bigcup_{U, V \in CSP\langle Q, S \rangle} C^+(U, V)$$

and it follows that there is a (unique) extension of the partial order relation $<$ from $C^+(Q, S)$ to $csp^+ \langle Q, S \rangle$. Together with the conclusion of (a) (above), this completes the proof of (i) and (ii).

(iii) If T passes through the event x, the result is an immediate consequence of the Intermediate Path Theorem (Th.31). The other possibility is where T meets

68

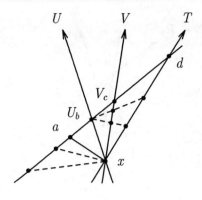

Figure 47

distinct paths U, V of $CSP\langle Q, S \rangle$ at events U_b, V_c. Then either $U_b < V_c$ or $U_b > V_c$ so we suppose, without loss of generality, that $x < U_b < V_c$ (Figure 47). Then $[x, V(U_b, \emptyset), V_c]$ so Theorem 27 implies that $T(x, \emptyset) < U_b$ (where $<$ is the induced linear ordering on T). Let a, d be any events of T such that $[T(x, \emptyset), a, U_b, V_c, d]$. Then Theorem 27 implies that $x < a < U_b < V_c < d$ in $\mathcal{C}(x, a, d)$ so that

$$\langle xa, U, V, xd \rangle .$$

Thus $a, d \in csp^+\langle U, V \rangle$ and obviously any event of T between U and V is also in $csp^+\langle U, V \rangle$. By the preceding Theorem 33, $CSP\langle U, V \rangle = CSP\langle Q, S \rangle$ so all events of T after x belong to $csp^+\langle Q, S \rangle$. \hfill q.e.d.

Theorem 35 (Containment)
Let Q, R, S be distinct paths which meet at distinct events b, c, d where $b = S \cap Q$, $c = Q \cap R$, $d = R \cap S$ such that $b < c < d$ (with respect to the partial order relation $<$ defined on $csp^+\langle Q, S \rangle$). Then
    (i) $csp^+\langle R, S \rangle \subset csp^+\langle Q, R \rangle \subset csp^+\langle Q, S \rangle$
   (ii) $\{x : \ x > d, \ x \in csp^+\langle Q, S \rangle\} \subseteq csp^+\langle R, S \rangle$

(see Figure 48).

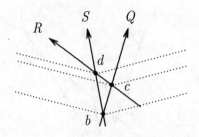

Figure 48

Proof Throughout this proof we will use the order relation $<$ on $csp^+\langle Q, S\rangle$.

(i) Take an event $e \in R$ with $c < d < e$ (Figure 49). By Theorem 16 there is an event $f \in Q$ such that $(b\ <)c < f < e$. Then the Second Collinearity Theorem (Th.7) implies that $bd(= S)$ meets (ef) at an event g such that $b < d < g$ and $f < g < e$. By the Second Existence Theorem (Th.14) there is an event $h \in Q$ with $h > g$ and a path $gh(= hg)$. The Second Collinearity Theorem implies that hg meets $ed(= R)$ at an event i such that $d < i < g$. The Coincidence Corollary (Th.32) implies that any path T of $CSP\langle R, S\rangle$ meets either $eg(= ef = gf)$ or $gi(= gh = hi)$ at an event after d. By the previous theorem (Th.34(iii)) this event of T belongs to $csp^+\langle R, S\rangle$ and $csp^+\langle Q, R\rangle$ and $csp^+\langle Q, S\rangle$. Again by the previous theorem (Th.34(iii)), $\{T_y\ :\ T_y > d,\ T_y \in T\}$ is contained in both $csp^+\langle Q, R\rangle$ and $csp^+\langle Q, S\rangle$. That is

$$csp^+\langle R, S\rangle \subset csp^+\langle Q, R\rangle \quad \text{and} \quad csp^+\langle R, S\rangle \subset csp^+\langle Q, S\rangle. \tag{1}$$

Again by Theorem 16, there is an event $j \in S$ such that $b < j < c$ and a path $W := jc$. The configuration S, W, Q with $b < j < c$ is analogous to Q, R, S with $b < c < d$ and the relation analogous to the second relation of (1) is

$$csp^+\langle W, Q\rangle \subset csp^+\langle S, Q\rangle. \tag{2}$$

The Ordered Coincidence Theorem (Th.30) implies that W meets gf at an event k such that $b < j < c < k < f < g(< e)$ whence $\langle R, Q, W\rangle$ and then (2) implies

$$csp^+\langle Q, R\rangle \subset csp^+\langle Q, S\rangle \tag{3}$$

which, together with the first relation of (1), completes the proof of (i).

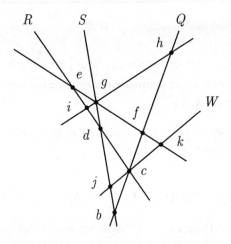

Figure 49

(ii) By the Ordered Coincidence Theorem any event of $csp^+\langle Q, S\rangle \setminus S$ after d belongs to a path (through d) which meets either gh or gf at some event after d. The result (ii) is now a consequence of Theorem 34(iii). $\qquad$ *q.e.d.*

5.5 The collinear set theorem

We define a "collinear set of events" in a manner which is analogous to the definition of a plane in absolute geometry. Let R, S be any two distinct paths which meet at some event. Then the *set of events collinear with R and S* is:

$$col[R, S] := \{W_x : W_x \in W, \text{ where } W \text{ meets } R \text{ and } S \text{ at distinct events } \}.$$

In the next theorem we will show that collinear sets of events have some properties which are analogous to the properties of coplanar sets of points in absolute geometry. However, it transpires that the collinear set of events has more structure than the geometrical plane for there is a relation of partial order on a collinear set. In part (iv) of the theorem we will show that for any event e and any path V (both in the collinear set) there is a *past set* $V(e, -)$ and a *future set* $V(e, +)$.

Theorem 36 (Collinear Set)

(i) Containment: *(the set of events of) any path which meets a collinear set in two distinct events, is contained in the collinear set. Accordingly we define the collinear set of paths*

$$COL[R, S] := \{U : U_x \text{ meets } col[R, S] \text{ in two distinct events}\}$$

and then

$$col[R, S] = \{U_x : U_x \in U \text{ where } U \in COL[R, S]\} .$$

(ii) Temporal Order Relation: *on each collinear set there is a partial order relation $<$. For events f, g such that $f < g$, there is a path fg.*

(iii) Collinearity: *if any three distinct paths (of a collinear set) meet at a single event, then one of the paths is between the other two.*

(iv) *For any path $V \in COL[R, S]$ and any event $e \in col[R, S]$ which does not belong to V, we define the past of e in V:*

$$V(e, -) := \{x : x < e, \ x \in V\}$$

and the future of e in V:

$$V(e, +) := \{x : x > e, \ x \in V\} .$$

Then the past and future of e in V are non–empty and

$$V(e, -) \ < \ V(e, \emptyset) \ < \ V(e, +)$$

where each event of V belongs to one of the three sets. The past and future sets are open with respect to the order topology and the unreachable set is closed.

(v) Uniqueness: *a given collinear set is uniquely determined by any two of its distinct paths which meet at a single event, or by any one of its paths and any one if its events which is not on the path.*

Proof (i)(a) As in the preceding definition, let R and S be distinct paths which meet at an event x_0. Take $y \in R$ with $y < x_0$ and take $x_1 \in S$ such that $[x_1, S(y, \emptyset), x_0]$ and then let $Q := x_1 y$. By the preceding theorem (Th.35(i)), $csp^+\langle R, S \rangle \subseteq csp^+\langle Q, S \rangle$. Let us denote $csp^+\langle R, S \rangle$ by $\mathcal{C}_0$ and $csp^+\langle Q, S \rangle$ by $\mathcal{C}_1$: then $\mathcal{C}_0 \subseteq \mathcal{C}_1$. Next we take an unbounded decreasing sequence

$$(x_n : x_n \in S, \ n \text{ a positive integer})$$

and let $Q^{(n)} := x_n y$ and $C_n := csp^+\langle Q^{(n)}, S\rangle$. Then by the preceding theorem,

$$C_0 \subseteq C_1 \subseteq C_2 \cdots \subseteq C_{n-1} \subseteq C_n$$

where the partial order relation $<$ has a unique extension from C_0 to C_n. Now we shall define

$$C := \cup_n C_n \qquad (1)$$

with the partial order relation $<$ on C defined by the preceding unique extensions. (It will transpire that $C = col[R, S]$).

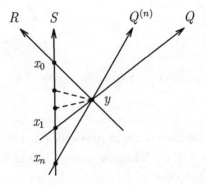

Figure 50

(b) We will now show that $C \subseteq col[R, S]$: first consider C_1. For any event $e \in C_1 = csp^+\langle Q, S\rangle$, there is a path $T = x_1 e \in CSP\langle Q, S\rangle$ such that either (α) $\langle T, S, Q\rangle$, (β) $\langle S, T, Q\rangle$ or (γ) $\langle T, Q, S\rangle$ (Figure 51).

For Case (α), the Second Existence Theorem (Th.14) and the Prolongation Theorem (Th.6) imply the existence of an event $z \in Q$ such that $z > e$, $z > x_0$ and such that z, e, x_0 do not belong to one path. Let w be the event at which the path ez meets S. There are two possibilities: if $w > x_0$ then the Ordered Coincidence Theorem (Th.30(i)) implies that R meets $ew(= wz)$ so that $e \in col[R, S]$; and if $w < x_0$, the same result is implied by the Third Collinearity Theorem (Th.15(i)). For Case (β) the Intermediate Path Theorem (Th.31) implies that T meets R at some event distinct from x_1 (and for Case (γ) the same result is implied by Theorem 32). In each case, $e \in col[R, S]$ and therefore $C_1 \subset col[R, S]$. Similar considerations apply to C_n with $x_n, Q^{(n)}$ taking the place of x_1, Q in the above argument; so for all n, it follows that $C_n \subseteq col[R, S]$ and hence

$$C \subseteq col[R, S] . \qquad (2)$$

73

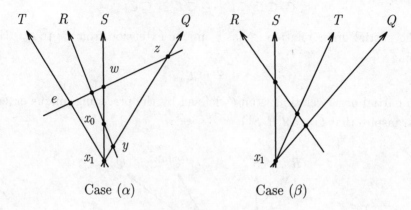

Case (α)   Case (β)

Figure 51

(c) We will now prove the containment property (i). Let T be a path which meets any two distinct events $a, b \in \mathcal{C}$. There is some $\mathcal{C}_m$ which contains both a and b. Theorem (Th.34(iii)) implies that

$$\{T_z : T_z > x_m, \ T_z \in T\} \subset \mathcal{C}_m .$$

Now for any event $T_c \in T$ such that $T_c < T(x_m, \emptyset)$, the Second Existence Theorem (Th.14) implies that there is some x_n (with $n > m$) such that $x_n < S(T_c, \emptyset)$ whence by Theorem 34(iii),

$$\{T_z : T_z > T_c, \ T_z \in T\} \subseteq \{T_z : T_z > x_n, \ T_z \in T\} \subseteq \mathcal{C}_n .$$

But T_c is arbitrary, hence

$$T \subseteq \bigcup_n \mathcal{C}_n = \mathcal{C} . \qquad (3)$$

In particular we have $R \subset \mathcal{C}$ and by definition $S \subset \mathcal{C}$ so for any path T which meets R and S at distinct events, the relation (3) applies whence $col[R, S] \subseteq \mathcal{C}$ which, together with the relation (2), implies that

$$\mathcal{C} = col[R, S] \qquad (4)$$

(thus establishing the parenthesized assertion which followed (1)). The relations (3) and (4) imply the Containment Property (i).

(ii) The temporal order relation $<$ on $\mathcal{C}$ ($= col[R, S]$) is specified in the definition (1) of $\mathcal{C}$. By Theorem 34 any pair of related events can be joined by a path which (by part (i) above) is contained in $col[R, S]$.

(iii) Let T, U, V be any three paths of $COL[R, S]$ which meet at an event z.

In the case where $z \notin S$, there exist $\mathcal{C}_m, \mathcal{C}_{m+1}$ which contain z and all events of T, U, V after z. Now the Containment Theorem (Th.35(i)) implies that all events of T, U, V after z belong to $csp^+ \langle x_m z, x_{m+1} z \rangle$ and Theorem 33 implies that $CSP \langle x_m z, x_{m+1} z \rangle$ is simply ordered. Thus one of the three paths T, U, V is between the other two.

In the other case where $z \in S$, the Density Theorem (Th.17) implies the existence of an event $y_m \in Q^{(m)}$ such that $x_m < y_m < z$. As in the preceding paragraph, the Containment Theorem (part(i)) implies that all events of T, U, V after z belong to $csp^+ \langle S, y_m z \rangle$ and $CSP \langle S, y_m z \rangle$ is simply ordered.

(iv) The path V meets some $\mathcal{C}_m$ in (at least) two distinct events, one of which we will call z. Thus there is a $\mathcal{C}_n$ which contains the event e and some ray of V, so $x_n < e$ and $x_n < z$. Hence there is another event $y \in V$ such that $y < x_n$. Thus $y < x_n < e$ so the event e can be connected to V by the path ye and hence, by the Second Existence Theorem, the past and future sets of e in V are non–empty.

(v) To demonstrate the first proposition of (v) we will show that, given any two distinct paths $T, U \in COL[R, S]$ such that T, U meet at some event z,

$$COL[T, U] = COL[R, S] .$$

(a) First we consider the case where $T, U \in CSP \langle R, S \rangle$; that is, $z = x_0$. Theorem 33 implies that

$$CSP \langle T, U \rangle = CSP \langle R, S \rangle \quad \text{and} \quad csp^+ \langle T, U \rangle = csp^+ \langle R, S \rangle . \tag{5}$$

For any path $V \in COL[R, S] \backslash CSP \langle R, S \rangle$, part (iv) implies that $V(x_0, +)$ exists and is contained in $csp^+ \langle R, S \rangle$ ($= csp^+ \langle T, U \rangle$), so by the Containment Property (part (i) of this theorem),

$$V \in COL[T, U] . \tag{6}$$

But V was arbitrary, hence (5) and (6) imply that

$$COL[R, S] \subseteq COL[T, U] \tag{7}$$

and the opposite containment follows by essentially the same argument with the pairs R, S and T, U interchanged, so that

$$COL[R, S] = COL[T, U] \quad \text{and} \quad col[R, S] = col[T, U] . \tag{8a}$$

(b) Second we consider the case where T, U meet at an event $z \neq x_0$. Part (iv) of this theorem implies that any path in $COL[R, S]$ meets either:

(α)   one path of $CSP\langle R, S \rangle$ at two distinct events after x_0, or

(β)   at least two distinct paths of $CSP\langle R, S \rangle$ at events after x_0.

Both cases can apply, separately, to T and U. The only case where both T and U satisfy condition (α), has already been considered in the preceding paragraph ((a) above). If T (or U or both) satisfy condition (β) then by the Intermediate Path Theorem (Th.31) and its Corollary (Th.32), T (or U or both) meets $CSP\langle R, S \rangle$ at events after x_0 in a set of paths which has the properties:

(A)   if it contains any two distinct paths of $CSP\langle R, S \rangle$, then it contains all paths between them (by Theorem 31)

(B)   if it contains a path, then it contains all paths on one side[1] of the path (by Theorem 32).

Take two paths V, W between T and U such that all four paths T, U, V, W are distinct: at least three of these paths will meet $CSP\langle R, S \rangle$ in sets of paths having the properties (A), (B) above (since otherwise we would have $T, U, V, W \in CSP\langle R, S \rangle$). The properties (A) and (B), which are analogous to the order properties of sub-intervals of the real numbers of the form $(-\infty, a)$ and (b, ∞), imply that at least two of these sets of paths have a non-trivial intersection (in the set-theoretic sense) containing two distinct paths (and all paths between them).

Thus there are (at least) two distinct paths R', S', of $CSP\langle R, S \rangle$ each of which meets two distinct paths T', U' of $CSP\langle T, U \rangle$ where T', U' are two of the four paths T, U, V, W. By the Containment Property (i) (above),

$$R', S' \in COL[T', U'] \quad \text{and} \quad T', U' \in COL[R', S'] .$$

By the result of paragraph (a) above

$$R', S' \in COL[T, U] \quad \text{and} \quad T', U' \in COL[R, S]$$

so by the Intermediate Path Theorem (Th.31) and its corollary (Theorem 32), together with the Containment Property (i) above,

$$R, S \in COL[T, U] \quad \text{and} \quad T, U \in COL[R, S] .$$

Now the Containment Property implies that

$$col[R, S] \subseteq col[T, U] \quad \text{and} \quad col[T, U] \subseteq col[R, S] \tag{8b}$$

which completes the proof of the first statement of (v). The second statement is a consequence of the first statement and (iv). $\hspace{2cm} q.e.d.$

The Collinear Set Theorem which has now been proved shows that there is a well-defined correspondence between a collinear set of paths $COL[\cdot, \cdot]$ and the corresponding collinear set of events $col[\cdot, \cdot]$. From now on the symbol Σ will be used to denote a collinear set of events: thus we will indicate that an event belongs to Σ by using the set membership symbol " $\in$ " and we will say that "a path is in Σ".

6.   Paths and optical lines in a collinear set

In this chapter we restrict our attention to a collinear set and develop the concept of "relative position" or directions "to the left" and "to the right" in Section 6.1. Forward and reverse signal functions are defined in Section 6.2 and it is shown that they are (weakly) order-preserving (Theorem 38), that they satisfy some "Signal Function Inequalities" (Theorem 39) and "Triangle Inequalities" (Theorem 40). This leads to Section 6.3 and the important "Signal" Theorem (Th.43) which states that forward and reverse signal functions are inverse of each other. In Section 6.4 we demonstrate the existence of "optical lines" which are sets of events which have the property that any pair are signal–related. "Optical lines" have a "sense of direction" which is either "to the left" or "to the right". At this stage we can not show that a maximal set of events with this property is contained in a collinear set, so we do not use the obvious physical description of a "light ray": the required containment property is eventually demonstrated, but this can be done no earlier than in Chapter 10 using the standard coordinate model of Minkowski space–time. Sections 6.5–6.7 are concerned with "reflections in paths", generalizations of the triangle inequalities, and a sufficient condition for the crossing of paths which is expressed in terms of signal functions. In Section 6.8 we define two types of "modified signal functions" which incorporate the "sense of direction" properties of left and right optical lines. This enables us to define "modified record functions" which are a first step towards the eventual coordinatization of a collinear set. In Section 6.9 we consider the properties of simultaneously coincident sets of collinear paths, or "collinear sub–SPRAYs". The chapter concludes with Section 6.10 where we state an existence theorem for paths in a collinear set.

We restrict our attention to a collinear set of events denoted by Σ, with a partial order relation $<$ and a corresponding set of paths: as with the concept of a plane in absolute geometry, we shall say that a path is *contained in* Σ, *belongs to* Σ or is *a path of* Σ. The concepts of *past*, *future* and *unreachable* sets are extended so that we now define

$$\begin{aligned}
\Sigma(v,+) &:= \{x : x > v,\ x \in \Sigma\} \\
\Sigma(v,-) &:= \{x : x < v,\ x \in \Sigma\} \\
\Sigma(v,\emptyset) &:= \{x : x \ngtr v \ \text{and}\ x \nless v,\ x \in \Sigma\} \ .
\end{aligned}$$

Events of the future (resp. past) set are the events *after* (resp. *before*) v (in Σ) .

6.1 The crossing theorem

The next theorem describes properties of a "sense of direction" expressed in terms of the two partial order relations "to the left of" and "to the right of" in a collinear set. The distinction between "left" and "right" is arbitrary and is specified as a convention for any given collinear set in order to facilitate its description.

Theorem 37 (Crossing)

A collinear set Σ can be given a sense of direction which has the following properties:

(i) *Each path T separates the events of $\Sigma \setminus T$ into two disjoint sets, called a* left *side of T and a* right *side of T.*

(ii) *A path W which has one event from the left side of T and one event from the right side of T meets T at an event v between the two given events: the path crosses T in the sense that the component of W before v is on one side of T while the component of W after v is on the other side.*

(iii) *The sense of direction (left to right) is consistent in the sense that:*

  (α) if (all events of) a path W is on the right side of a path T, then the right side of W is contained in the right side of T (and the left side of T is contained in the left side of W).

  (β) if W crosses T at an event v and the events of W after (resp. before) v are on the right side of T, then the right side of W after (resp. before) v is contained in the right side of T after (resp. before) v.

(iv) *If two paths meet, then they cross.*

Remarks   1. The statement of the theorem contains the definitions of *left side, right side* and *sense of direction*.

2. Part (iii) has a corresponding result with the words "left" and "right" interchanged.

3. The *right side of T after v* is the intersection of the right side of T with $\Sigma(v,+)$. A similar definition applies to sides before v.

Proof (a) We will first show that the result applies to the events of a compact collinear set. Let the paths Q, S and the events x, Q_m, S_n be defined as in the proof of the Forward Collinear sub–spray Theorem (Th.34). The Compact Collinear Set Theorem (Th.18) implies that any path T which meets $\mathcal{C}(x, Q_m, S_n)$ in two distinct events, also meets the boundary in two distinct events which we shall denote as T_0, T_1 where $T_0 < T_1$. These events separate the remaining subset of $\mathcal{B}(x, Q_m, S_n)$ into two disjoint open components which we shall call the *left component* $\mathcal{LB}$ and the *right component* $\mathcal{RB}$ and which are defined such that (Figure 52): if $[x\ T_0\ Q_m]$ or if $[Q_m\ T_0\ S_n]$ then $x \in \mathcal{RB}$, if $[x\ T_0\ S_n]$ then $x \in \mathcal{LB}$ while if $T_0 = x$ then $Q_m \in \mathcal{LB}$ and $S_n \in \mathcal{RB}$: each component is connected. We next define the *left side of T in* $\mathcal{C}(x, Q_m, S_n)$ to be

$$\mathcal{L} := \{l :\ [l_1\ l\ l_2];\ l_1, l_2 \in \mathcal{LB}\}$$

and the *right side of T in* $\mathcal{C}(x, Q_m, S_n)$ to be

$$\mathcal{R} := \{r :\ [r_1\ r\ r_2];\ r_1, r_2 \in \mathcal{RB}\}\ .$$

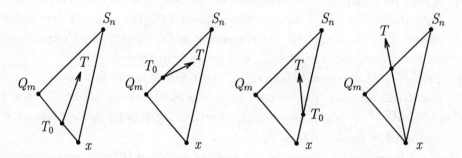

Figure 52

Since each component of the boundary is connected the Third Collinearity Theorem (Th.15(iii)) implies that $\mathcal{L}$ and $\mathcal{R}$ are disjoint sets.

Next we define the *left boundary* $\overline{\mathcal{LB}} := \mathcal{LB} \cup \{T_0, T_1\}$ and the *right boundary* $\overline{\mathcal{RB}} := \mathcal{RB} \cup \{T_0, T_1\}$. We will show that (for $i = 0$ *or* 1 and $l \in \mathcal{LB}$),

$$[T_i\ e\ l] \Longrightarrow e \in \mathcal{L}\ . \tag{1a}$$

Let j be 1 if i is 0, or 0 if i is 1. Since there is no fastest path (Theorem 16), there is an event l_3 between l and T_j on the boundary $\mathcal{B}(x, Q_m, S_n)$ and a path $l_3 e$. The

Compact Collinear Set Theorem (Th.18) implies that l_3e meets the boundary in a second event: by the Third Collinearity Theorem (Th.15(iii)) this event is between T_i and l on the boundary, that is in $\mathcal{LB}$. By definition, the event e belongs to $\mathcal{L}$. Corresponding to (1a), a similar result (1b) applies for an event $r \in \mathcal{RB}$.

For any event e in the interior of $\mathcal{C}(x, Q_m, S_n)$ there is a path xe (as in the Forward Collinear sub–spray Theorem (Th.34)) which, by the Compact Collinear Set Theorem 18 meets $(Q_m S_n)$ in an event f. We will now proceed to show that if $e \notin (T_0 T_1)$ then either $e \in \mathcal{L}$ or $e \in \mathcal{R}$. Firstly if xe does not meet $(T_0 T_1)$ then both x and f belong to the same boundary component and so, by definition of $\mathcal{L}$ and $\mathcal{R}$, the event e belongs to one or to the other. Secondly if xe meets $(T_0 T_1)$ in an event g then $T_0 < g < T_1$ and either $e < g < f$ or $g < e < f$ and, accordingly, either $e < T_1$ or $T_0 < e$ so, accordingly there is a path eT_1 or there is a path $T_0 e$. Now the Compact Collinear Set Theorem (Th.18) and the Axiom of Uniqueness of Paths (Axiom I3) imply that this path meets one of the boundary components $\mathcal{LB}$ or $\mathcal{RB}$ and, by (1a,b), this implies that e belongs to $\mathcal{L}$ or to $\mathcal{R}$. This establishes a result analogous to (i) for the compact collinear set $\mathcal{C}(x, Q_m, S_n)$.

Now consider a path which contains an event $l \in \mathcal{L}$ and an event $r \in \mathcal{R}$. The Compact Collinear Set Theorem implies that the path lr meets $\mathcal{B}(x, Q_m, S_n)$ in two distinct events and one event must be in $\mathcal{LB}$ while the other event must be in $\mathcal{RB}$ since otherwise the pair of events l, r would both be in $\mathcal{L}$ or would both be in $\mathcal{R}$. Now the Third Collinearity Theorem (Th.15(iii)) implies that lr crosses T at some event v in $\mathcal{C}(x, Q_m, S_n)$ and by the previous paragraph, all events before v are on one side of T while all events after T are on the other side of T; we have now demonstrated the analogue of part (ii) for the compact collinear set.

A result analogous to (iii)(α) is easily obtained by denoting the left component and right component for the path W' as $\mathcal{LB}'$ and $\mathcal{RB}'$ respectively. Then $\mathcal{LB} \subset \mathcal{LB}'$ and $\mathcal{RB} \supset \mathcal{RB}'$ from which it follows immediately that $\mathcal{L} \subset \mathcal{L}'$ and $\mathcal{R} \supset \mathcal{R}'$. A result analogous to (iii) (β) for paths which meet at v in $\mathcal{C}(x, Q_m, S_n)$ is obtained in a similar way.

The result analogous to (iv) for the compact collinear set $\mathcal{C}(x, Q_m, S_n)$ is an immediate consequence of the Third Collinearity Theorem (Th.15(iii)).

(b) If we now consider sequences of events $(Q_m : m = 1, 2, \cdots)$ and $(S_n : n = 1, 2, \cdots)$ as in the proof of the Forward Collinear sub–spray Theorem (Th.34) results analogous to (i),(ii),(iii) (α),(iv) are seen to apply, firstly, to $\mathcal{C}^+(Q, S)$ (as defined in the proof of Theorem 34) and, secondly, to $csp^+\langle Q, S \rangle$. The result (iii) ($\beta$) applies to two paths which meet at x, for events after x.

(c) As in the proof of the Collinear Set Theorem (Th.36), consider a sequence of events $(x_n : n = 1, 2, \cdots ; x_n \in S)$, a sequence of paths $(Q^{(n)})$ and a corresponding sequence of forward collinear sub–sprays (C_n). Then results analogous to (i),(ii),(iii) (α),(iv) apply to each C_n and hence to their union $col[Q, S]$.

(d) The result (iii) (β) has already been proved (in (a) above) for two paths which meet at v and for events after v. Since the direction of temporal order is merely a convention, the analogous result applies to the set of events which precede the event of coincidence. $q.e.d.$

It is convenient and useful to have a notation which denotes relative positions or betweenness of paths and, for this purpose we extend the application of the symbols $\langle \cdots \rangle$ for sets of paths which need not be simultaneously coincident. Thus if R is on one side of Q after an event x and if S is on the same side of Q and also on the same side of R after x we say that R is *between Q and S after x* and write

$$\langle Q, R, S \rangle \text{ after } x .$$

A similar definition applies for *betweenness before* an event. For paths Q, R, S which never meet and therefore never cross, we may write simply $\langle Q, R, S \rangle$.

6.2 Order properties of signal functions

The "signal functions" alluded to in Section 5.1 will now be defined in relation to the direction of time specified by the temporal order relation $<$ on the collinear set Σ. In this section we will obtain results concerning order properties of signal functions and we will also establish an existence theorem (Theorem 42). The Axiom of Continuity (Axiom C) asserts that each path has a completeness property which ensures the existence of a sufficient number of events to allow for the definition of "signal functions". These functions may be interpreted in physical terms as correspondences defined by "fastest signals" or "light signals" (Figure 53).

We define two types of signal function. The *forward signal function* is defined to be

$$f^+_{RQ} : Q \;\rightarrow\; R$$

$$Q_x \;\mapsto\; \begin{cases} \inf R(Q_x, +), & \text{if } Q \text{ and } R \text{ do not meet at } Q_x \\ Q_x, & \text{if } Q \text{ and } R \text{ meet at } Q_x \end{cases}$$

and the *reverse signal function* is defined to be

$$f^-_{RQ} : Q \rightarrow R$$

$$Q_x \mapsto \begin{cases} \sup R(Q_x, -), & \text{if } Q \text{ and } R \text{ do not meet at } Q_x \\ Q_x, & \text{if } Q \text{ and } R \text{ meet at } Q_x \end{cases}$$

where the existence of $\inf R(Q_x, +)$ and $\sup R(Q_x, -)$ is ensured by the Continuity Theorem (Th.12). Actually these events are (respectively) the minimum and maximum of the unreachable set $R(Q_x, \emptyset)$ as demonstrated in Theorem 16. These signal functions will eventually turn out to correspond physically to "light signals" and, as might be expected, it will transpire (in the Signal Theorem (Th.43)) that pairs of functions, such as f^+_{RQ} and f^-_{QR} are inverses of each other. However this is not immediately apparent within the present axiomatic system since there is no axiom corresponding to the property which Kronheimer and Penrose (1967) and Woodhouse (1973) have called "future and past distinguishing". The next theorem supersedes Theorem 28 of Section 5.1.

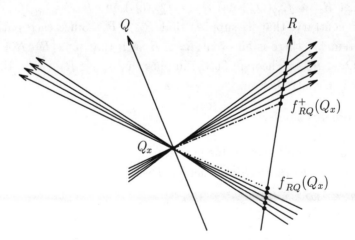

Figure 53   Signal functions are defined by considering limiting properties of sets of paths: forward signal functions are indicated by broken lines with short and long dashes and reverse signal functions are indicated by broken lines with bunched dots

Theorem 38 (Signal Functions are Weakly Order-Preserving)
Given two paths Q, R in Σ with events $Q_x, Q_z \in Q$ then

(i) $Q_x < Q_z \implies f_{RQ}^+(Q_x) \leq f_{RQ}^+(Q_z)$

(ii) $Q_x < Q_z \implies f_{RQ}^-(Q_x) \leq f_{RQ}^-(Q_z)$.

Remark The paths Q, R need not meet as required in Theorem 28.

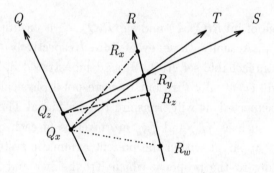

Figure 54

Proof (i) Let $R_x := f_{RQ}^+(Q_x)$ and $R_z := f_{RQ}^+(Q_z)$ and $R_w := f_{RQ}^-(Q_x)$ (Figure 54). Suppose the contrary; that is, suppose that $R_x > R_z$. Since each path is dense in itself (Theorem 17) there is an event $R_y \in R$ such that $\max\{R_w, R_z\} < R_y < R_x$. Now $Q_x < Q_z < R_y$ so Theorem 36 (ii) implies that $Q_x < R_y$ and so there is a path $Q_x R_y$ which is a contradiction.

(ii) The proof is similar.   *q.e.d.*

Theorem 39 (Signal Function Inequalities)
Let Q, R be two paths in a collinear set Σ. Then

(i) $f_{QR}^- \circ f_{RQ}^+ \leq i_{QQ}$
(ii) $f_{QR}^+ \circ f_{RQ}^- \geq i_{QQ}$

where i_{QQ} is the identity function on Q. Now let $R_c := f_{RQ}^+(Q_a), \; Q_b := f_{QR}^-(R_d)$.
Then

(iii) $Q_a < Q_b \iff R_c < R_d$
(iv) $R_c > R_d \implies Q_a \geq Q_b$.

Remark See Figure 55.

84

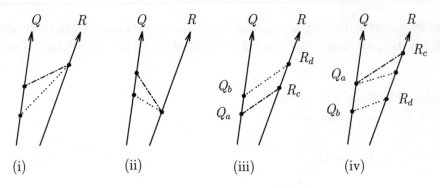

(i)   (ii)   (iii)   (iv)

Figure 55   The Signal Function Inequalities

Proof (i) Given any event $Q_w \in Q$, let $R_y := f_{RQ}^+(Q_w)$ and let $Q_x := f_{QR}^-(R_y)$. We suppose the contrary; that is, we suppose that $Q_w < Q_x$ (Figure 56a) and obtain a contradiction. Since $Q_x = f_{QR}^-(R_y)$, the supposition $Q_w < Q_x$ implies the existence of a path joining Q_w and R_y. But then $R(Q_w, +)$ has a first event which contradicts Theorem 16.

(ii) A similar, though time-reversed, proof applies.

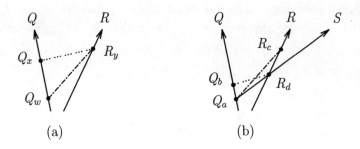

(a)   (b)

Figure 56

(iii) We first show that $Q_a < Q_b \implies R_c < R_d$. By (i), $R_c \neq R_d$, so either (a) $R_c < R_d$ or (b) $R_c > R_d$. We shall show that (b) (Figure 56b) leads to a contradiction. Since $Q_a < Q_b = f_{QR}^-(R_d)$ it follows that $Q_a < R_d$ which, with the supposition $R_d < R_c$, implies that $Q_a < R_c$ and there is a path joining Q_a and R_c which contradicts Theorem 16: Case (b) is therefore excluded, leaving Case (a) as the only remaining possibility. The reverse implication follows in the same way since the forward direction of time is merely a convention.

85

(iv) We suppose the contrary; namely that $Q_b > Q_a$ which, with $R_c > R_d$ implies that $R_c > R_d > Q_a$ which is a contradiction, since there is no path joining Q_a and R_c. *q.e.d.*

Theorem 40 (Triangle Inequalities)
Let Q, R, S be three paths in Σ. Then

(i) $\qquad f_{SR}^+ \circ f_{RQ}^+ \geq f_{SQ}^+$

(ii) $\qquad f_{SR}^- \circ f_{RQ}^- \leq f_{SQ}^-$

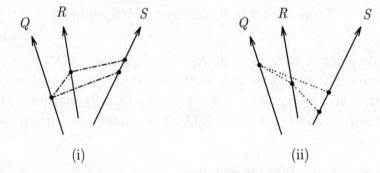

(i) $\qquad\qquad\qquad\qquad\qquad$ (ii)

Figure 57 $\quad$ The Triangle Inequalities

Proof (i) (See Figure 58). Take any event $Q_u \in Q$ and let $R_v := f_{RQ}^+(Q_u)$ and $S_w := f_{SR}^+(R_v)$. Consider any event $S_z \in S$ such that $S_z > S_w$; we will show that $Q_u < S_z$.

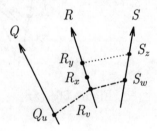

Figure 58

Let $R_y := f_{RS}^-(S_z)$. By the Signal Function Inequalities (Th.39(iii)), $R_y > R_v$. Theorem 16 implies the existence of an event $R_x \in R$ such that $R_v < R_x < R_y$. Therefore $Q_u < R_x$ and $R_x < S_z$ whence $Q_u < S_z$.

(ii) A similar proof applies. $\hfill$ *q.e.d.*

Given two events a, b and paths Q, R such that $f_{RQ}^+(a) = b$ we say that a and b are *(forward) signal–related* and write $a \, \sigma \, b$. The binary relation σ is called the *signal relation* and an expression of the form $a \, \sigma \, b$ may be read as " a signal goes from a to b " or " a signal leaves a and arrives at b ". (We do not use the symbol σ to apply to reverse signal functions.) Given three events a, b, c such that $a \, \sigma \, b$ and $b \, \sigma \, c$ and $a \, \sigma \, c$, we say that the ordered set of events is *in optical line* (a more detailed definition is given prior to the statement of Theorem 44). A similar definition applies to a set of three events which are related by the reverse signal function.

Theorem 41 (Second Optical Line Theorem)
Let Q, R, S meet at x. If $\langle \, Q, R, S \, \rangle$ then, for all $Q_u \in Q$,
(i)   *the events $Q_u, f_{RQ}^+(Q_u), f_{SQ}^+(Q_u)$ are in optical line*
(ii) *the events $Q_u, f_{RQ}^-(Q_u), f_{SR}^-(Q_u)$ are in optical line .*

Figure 59   The Second Optical Line Theorem

Proof   The result is trivial for the event x at which Q, R, S coincide. For any event Q_u at which the three paths do not meet, the Second Existence Theorem (Th.14) implies the existence of an event S_z such that $[x, S(Q_u, \emptyset), S_z]$ and then the previous Optical Line Theorem (Th.19) applies.

Theorem 42 (Third Existence Theorem)
Let Q, U be distinct paths which meet at some event x and let $Q_a, Q_b \in Q$ be distinct events such that (Figure 60)

$$x < Q_a < Q_b \quad \text{and} \quad f_{UQ}^+(Q_a) > f_{UQ}^-(Q_b) \, .$$

Then there exists a unique path S between Q and U such that

$$f_{SQ}^+(Q_a) \; = \; f_{SQ}^-(Q_b) \, .$$

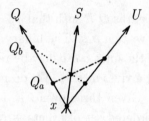

Figure 60   The Third Existence Theorem

Proof   (a) By the Intermediate Path Theorem (Th.31) there is a set of paths between Q and U which we can partition into two disjoint subsets $\mathcal{L}$ and $\mathcal{R}$, called left- and right-subsets respectively such that:

$$\mathcal{L} := \{R: f_{RQ}^+(Q_a) < f_{RQ}^-(Q_b), \langle Q, R, U\rangle\} \tag{1a}$$

$$\mathcal{R} := \{T: f_{TQ}^+(Q_a) \geq f_{TQ}^-(Q_b), \langle Q, T, U\rangle\} \tag{1b}$$

(b) We will first show that $\mathcal{L}$ and $\mathcal{R}$ are non–empty sets. Let $U_y := f_{UQ}^-(Q_b)$ and $U_z := f_{UQ}^+(Q_a)$ and take a path W through Q_a and an event of U after U_z (Figure 61). By Theorem 16 there is an event $e \in W$ such that $Q_a < e < Q_b$ and by the Intermediate Path Theorem (Th.31) there is a path R through x and e. Thus by the previous theorem

$$f_{RQ}^+(Q_a) < e < f_{RQ}^-(Q_b)$$

whence $R \in \mathcal{L}$ and so $\mathcal{L}$ is not empty. Similarly, we can demonstrate the existence of a path T such that $\langle Q, T, U\rangle$ and

$$f_{TU}^+(U_y) < f_{TU}^-(U_z) .$$

Now by the previous theorem and the Signal Function Inequalities (Th.39 (i,ii)),

$$\begin{aligned}
f_{TU}^-(U_z) &= f_{TU}^- \circ f_{UT}^+ \circ f_{TQ}^+(Q_a) \leq f_{TQ}^+(Q_a)\\
f_{TU}^+(U_y) &= f_{TU}^+ \circ f_{UT}^- \circ f_{TQ}^-(Q_b) \geq f_{TQ}^-(Q_b)
\end{aligned}$$

which, combined with the preceding inequality, imply that

$$f_{TQ}^-(Q_b) \leq f_{TU}^+(U_y) < f_{TU}^-(U_z) \leq f_{TQ}^+(Q_a)$$

so that $T \in \mathcal{R}$ and $\mathcal{R}$ is not empty.

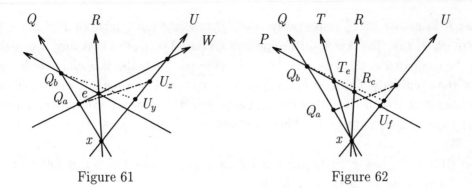

Figure 61   Figure 62

(c) Next we show that any paths $R \in \mathcal{L}$ and $T \in \mathcal{R}$ have the ordering $\langle Q, R, T, U \rangle$. Take an event $R_c \in R$ such that

$$Q_a \; < \; R_c \; < \; Q_b \tag{2}$$

and a path P through R_c and Q_b. By the Intermediate Path Theorem (Th.31) or the Coincidence Corollary (Th.32), P meets T at an event T_e and U at an event U_f such that either:

$$\text{(i)} \quad x < U_f < T_e < R_c < Q_b \quad \text{or (ii)} \quad x < U_f < R_c < T_e < Q_b \; . \tag{3}$$

In the Case (ii) (Figure 62), the inequalities (2) and (3(ii)) imply that

$$f_{TR}^+ \circ f_{RQ}^+(Q_a) \; < \; T_e$$

and by the Triangle Inequality (Th.40)

$$f_{TQ}^+(Q_a) \; < \; T_e \; . \tag{4}$$

Also, the inequality (3(ii)) implies that

$$f_{TQ}^-(Q_b) \; > \; T_e \; . \tag{5}$$

Now the inequalities (4) and (5) imply that $T \in \mathcal{L}$, which is a contradiction. Hence Case (ii) can not occur and we are left with Case (i) for which Theorems 31 and 33 imply that $\langle Q, R, T, U \rangle$, whence

$$R \in \mathcal{L} \; \text{ and } \; T \in \mathcal{R} \; \Longrightarrow \; \langle Q, R, T, U \rangle \; . \tag{6}$$

(d) We can order the set of paths between Q and U as follows. Consider *any* two paths $R, T \in CSP\langle Q, U \rangle$ (with R, T not necessarily in $\mathcal{L}, \mathcal{R}$ respectively) which

89

meet P at events R_c, T_d respectively: then R is to the left (right) of T if $R_c > T_d$ (resp. $R_c < T_d$). The Axiom of Uniqueness (Axiom I3) implies that any two paths may be compared in this way, and Theorem 33 implies that this is a simple ordering. Now the Continuity Theorem (Th.12) implies that there is a path S which belongs to either $\mathcal{L}$ or $\mathcal{R}$ and which has the property that it is between any (other) path of $\mathcal{L}$ and any (other) path of $\mathcal{R}$. There are now two distinct cases: (i) $S \in \mathcal{L}$ or (ii) $S \in \mathcal{R}$.

Case (i) $S \in \mathcal{L}$. Let $a' := f^+_{SQ}(Q_a)$ and $b' := f^-_{SQ}(Q_b)$. Since $S \in \mathcal{L}$ it follows that $a' < b'$ and the previous theorem implies that

$$f^+_{US}(a') = U_z \quad \text{and} \quad f^-_{US}(b') = U_y$$

so that the paths S, U and the events a', $b' \in S$ satisfy the conditions of this theorem as originally stated for the paths Q, U and the events Q_a, Q_b. Thus by parts (a) and (b) of this proof there is a path R in the corresponding left side $\mathcal{L}'$ such that $\langle Q, S, R, U \rangle$ and the previous theorem implies that

$$f^+_{RQ}(Q_a) = f^+_{RS}(a') < f^-_{RS}(b') = f^-_{RQ}(Q_b)$$

so that $R \in \mathcal{L}$, which is a contradiction.

Case (ii) $S \in \mathcal{R}$. If $f^+_{SQ}(Q_a) > f^-_{SQ}(Q_b)$, a similar argument shows that there is a path in $\mathcal{R}$ to the left of S, which is also a contradiction. The only remaining possibility is that $S \in \mathcal{R}$ and that

$$f^+_{SQ}(Q_a) = f^-_{SQ}(Q_b) \; .$$

(e) In order to prove uniqueness we consider the possibility that there are two paths S and S^* such that either $\langle Q, S, S^*, U \rangle$ or $S = S^*$. We let $S_a := f^+_{SQ}(Q_a)$, $S_b := f^-_{SQ}(Q_b)$ and let S^*_a, S^*_b be defined similarly. Then

$$S_a = S_b \quad \text{and} \quad S^*_a = S^*_b$$

and the previous theorem implies that

$$f^+_{S^*S}(S_a) = S^*_a \quad \text{and} \quad f^-_{S^*S}(S_b) = S^*_b$$

whence combining all four equalities we see that

$$f^+_{S^*S}(S_a) = f^-_{S^*S}(S_b) = f^-_{S^*S}(S_a) \; .$$

This equality would contradict Axiom I5 if S and S^* were distinct, so we conclude that S is unique. $\qquad\qquad q.e.d.$

6.3 The signal theorem

This theorem establishes properties of signal functions which were taken as axiomatic in the non-independent axiomatic systems of Walker (1948, 1959) and Schutz (1973).

Theorem 43 (Signal)
Let Q, R be any two paths in a collinear set Σ.

(i) *Signal functions have inverses and, for any paths Q and R,*

$$f^-_{RQ} = (f^+_{QR})^{-1} \quad \text{and} \quad f^+_{RQ} = (f^-_{QR})^{-1} .$$

(ii) *Signal functions are one-to-one onto mappings.*

(iii) *Signal functions are strictly order-preserving.*

(iv) *Signal functions are continuous (with respect to the order topology on each path).*

(v) *If Q' is a subset of Q and Q' is bounded above, then it has a supremum and*

$$\sup\{f^{\pm}_{RQ}(Q_t) : Q_t \in Q'\} = f^{\pm}_{RQ}(\sup\{Q_t : Q_t \in Q'\}) .$$

(A similar result applies if Q' is bounded below).
In particular if $(Q_n : n = 1, 2, \cdots ; Q_n \in Q)$ is a bounded monotone sequence, then it has a limit and

$$\lim_{n \to \infty} f^{\pm}_{RQ}(Q_n) = f^{\pm}_{RQ}(\lim_{n \to \infty} Q_n) .$$

Remarks 1. Thus both forward and reverse signal functions may be expressed in terms of the (forward) signal function

$$F_{RQ} := f^+_{RQ}$$

which, as we will see in Chapter 10, corresponds physically to a "light signal". The reverse signal function is related to the inverse of the (forward) signal function although the order of the path subscripts must be reversed, thus:

$$F^-_{QR} = F^{-1}_{RQ} .$$

2. In Section 6.8 and thereafter, superscript $+$ and $-$ signs will be used to indicate "modified signal functions" as in Walker (1948, 1959) and Schutz (1973, 1981). After the proof of the present theorem, we will no longer use the concepts of "forward" and "reverse" signal functions since they can be replaced by the signal function or its inverse.

Proof (see Figure 63). (a) We first consider the special case of two distinct paths Q, U which meet at an event x and we consider any event $e \in U$ such that $e > x$. Let $h := f_{QU}^-(e)$ and let $b := f_{UQ}^+(h)$: the Signal Function Inequalities (Th.39) imply that $b \geq e$. In order to show that $b = e$ we will suppose the contrary; that is, we will suppose that $b > e$, and then deduce a contradiction. Let $a := f_{QU}^+(b)$; then by the Signal Function Inequalities

$$(\alpha)\ f_{UQ}^-(a) = b \quad \text{or} \quad (\beta)\ f_{UQ}^-(a) < b\ .$$

These cases will be considered separately.

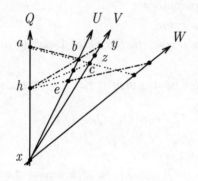

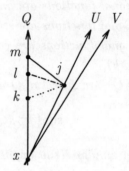

Figure 63   Figure 64

Case (α) $f_{UQ}^-(a) = b$. Since there is no fastest path (Theorem 16), there is a path W such that $\langle Q, U, W \rangle$. Now if $f_{WU}^-(b) \leq f_{WU}^+(e)$ then paragraph (b) of the proof of the preceding theorem (Theorem 42), (and with U, V, W taking the place of Q, R, U) implies the existence of a path V such that $\langle Q, U, V, W \rangle$ and such that

$$f_{VU}^-(b) > f_{VU}^+(e)\ ;$$

while if

$$f_{WU}^-(b) > f_{WU}^+(e)$$

then we simply let $V := W$.

Now let $c := f_{VU}^-(b)$, then

$$c = f_{VU}^-(b) > f_{VU}^+(e)\ ,$$

whence by the Triangle Inequality (Th.40(ii)) and the Signal Function Inequality (Th.39(iii))

$$f_{UQ}^+(h) = b \geq f_{UV}^-(c) > e\ .$$

By the Signal Function Inequality (Th.39(i)) and the weak order-preserving property of signal functions (Theorem 38) these relations imply that

$$h \geq f_{QU}^- \circ f_{UQ}^+(h) = f_{QU}^-(b) \geq f_{QU}^- \circ f_{UV}^-(c) \geq f_{QU}^-(e) = h$$

(whence all these inequalities can be replaced by equalities). Also the Second Optical Line Theorem (Th.41) together with the above equalities, implies that

$$f_{VQ}^-(a) = f_{VU}^- \circ f_{UQ}^-(a) = f_{VU}^-(b) = c \quad \text{and} \quad f_{QV}^-(c) = h$$

so that

$$f_{QU}^- \circ f_{UQ}^-(a) = h \quad \text{and} \quad f_{QV}^- \circ f_{VQ}^-(a) = h .$$

Since signal functions are weakly order-preserving (Theorem 38) these equalities imply that

$$Q(a, U, x, \emptyset) = Q(a, V, x, \emptyset)$$

so the Axiom of Isotropy (Axiom S) and Theorem 21, together with the Collinear Set Theorem (Th.36), imply that there is a bijection

$$\theta : \ col[Q, U] \longrightarrow col[Q, U]$$

such that $\theta(U) = V$ and such that all events of Q are invariant. (Actually the bijection θ is the restriction of an isotropy mapping to $col[Q, U] = \Sigma$).

We will show that θ maps signals from U to Q onto signals from V to Q. Consider any event $j \in U$ with $j > x$ (Figure 64) and let

$$k := f_{QU}^-(j) \quad \text{and} \quad l := f_{QU}^+(j) .$$

Now $j \in U$ so $\theta(j) \in \theta(U)$; furthermore, for each event $m \in Q$ with $m > l$ there is a path jm which is mapped bijectively by θ onto a path through $\theta(j)$ and $\theta(m)$ (which is just m); thus

$$f_{Q\theta(U)}^+(\theta(j)) = f_{QU}^+(j) = l \tag{1a}$$

and similarly

$$f_{Q\theta(U)}^-(\theta(j)) = f_{QU}^-(j) = k \tag{1b}$$

We have thus shown that the mapping θ sends signals from U to Q onto signals from V to Q. A similar argument applied to an arbitrary event $z \in V$ and with $\psi := \theta^{-1}$ shows that

$$f_{Q\psi(V)}^\pm(\psi(z)) = f_{QV}^\pm(z) . \tag{2}$$

Now let $y := f^+_{VU}(b)$; then for any $z \in V$ such that $c \leq z \leq y$, the Signal Function Inequality (Th.39(iv)) and the weak order-preserving property of signal functions imply that

$$f^-_{QV}(z) = h \quad \text{and} \quad f^+_{QV}(z) \geq a$$

and, by equation (2),

$$f^-_{Q\psi(V)}(\psi(z)) = f^-_{QU}(\psi(z)) = h$$

so by the Signal Function Inequalities (Th.39(iv))

$$\psi(z) \leq b \tag{3a}$$

Also

$$\begin{aligned} f^+_{QU}(\psi(z)) &= f^+_{Q\psi(V)}(\psi(z)) \\ &= f^+_{QV}(z) \qquad , \quad \text{by (2)} \\ &\geq a \end{aligned}$$

by the Signal Function Inequality 39(iv) since $f^-_{VQ}(a) = c$. Thus, by the weak order-preserving property,

$$\psi(z) \geq b \ . \tag{3b}$$

Combining (3a) and (3b) we obtain

$$\psi(z) = b \ . \tag{3}$$

However z was arbitrary, so

$$\psi : \{z : c \leq z \leq y, \ z \in V\} \longrightarrow \{b\} \ ,$$

but this contradicts Theorem 21 which requires that $\psi : col[Q, U] \longrightarrow col[Q, U]$ is a bijection.

Case (β) $f^-_{UQ}(a) < b$. We will show that this case implies the existence of a configuration of paths and events with the same order and signal relations which specified Case (α): this will then imply a contradiction.

Let the paths between Q and U be placed in two disjoint sets

$$\mathcal{L} = \{P : f^-_{PU}(b) = f^-_{PU}(e), \ \langle Q, P, U \rangle \text{ or } P = Q\}$$
$$\mathcal{R} = \{P : f^-_{PU}(b) > f^-_{PU}(e), \ \langle Q, P, U \rangle \text{ or } P = U\} \ .$$

clearly $Q \in \mathcal{L}$ and $U \in \mathcal{R}$ so both sets are non–empty and the Second Optical Line Theorem (Th.41) implies that no path of $\mathcal{R}$ can lie between any two paths of $\mathcal{L}$. We will now demonstrate the existence of a path $S \in \mathcal{L}$ which is between every other path of $\mathcal{L}$ and every path of $\mathcal{R}$.

First we will show that the set $\mathcal{R}$ is "open on the left" with respect to the order topology induced by the simple ordering of $CSP\langle Q, U \rangle$ (Theorem 33). By Axiom I5, $f_{QU}^+(e) > h$ and $Q \in \mathcal{L}$ so $f_{QU}^-(b) = h$, whence the Third Existence Theorem (Th.42) implies the existence of a path R between Q and U such that (Figure 65)

$$f_{RU}^+(e) = f_{RU}^-(b) \ . \tag{4}$$

If we let $b' := f_{RU}^-(b)$ and $e' := f_{RU}^-(e)$, then Axiom I5 and equation (4) imply that $b' > e'$ and so the set $\mathcal{R}$ has at least one path (namely R) other then U. Now for any path $V \in \mathcal{R}$ with $b'' := f_{VU}^-(b)$ and $e'' := f_{VU}^-(e)$ we can define sets $\mathcal{L}'$ and $\mathcal{R}'$ similarly and so (using a similar argument) $\mathcal{R}'$ has at least one path other than V. Furthermore, for any path $W \in \mathcal{R}'$,

$$f_{WV}^-(b'') > f_{WV}^-(e'') \quad \text{and} \quad f_{WU}^- = f_{WV}^- \circ f_{VU}^-$$

so

$$f_{WU}^-(b) > f_{WU}^-(e) \ ;$$

that is, $W \in \mathcal{R}$ and so $\mathcal{R}$ has no "left-most path". Second, the Intermediate Path Theorem (Th.31) and Theorem 33 imply that the set of paths $\mathcal{L} \cup \mathcal{R}$ has the same properties of order and completeness as the set of events on a path, so the set $\mathcal{L}$ must have a "right-most path" S (which could be Q). Now let $h^* := f_{SU}^-(b)$ and let $h'' := f_{SQ}^+(h)$. Then b, h^*, h are in reverse optical line and, by the definition of b, the events h, h'', b are in forward optical line. Now the Signal Function Inequalities (Th.39(ii)) applied to the triple of events h^*, h, h'' implies that $h^* \leq h''$ and also implies that $f_{US}^+(h^*) \geq b$ while the weak order-preserving property of signal functions (Theorem 38) implies that $f_{US}^+(h^*) \leq b$. Thus $f_{US}^+(h^*) = b$.

Now by analogy with the considerations prior to Case (α), we let $a^* := f_{SU}^+(b)$ (Figure 66). By the Third Existence Theorem (Th.42), there is a path $T \in \mathcal{R}$ such that $\langle Q,S,T,U \rangle$ or $T = U$ with

$$f_{TS}^-(a^*) = f_{TS}^+(h^*) =: b^*$$

so that h^*, b^*, b are in forward optical line. Let $e^* := f_{TU}^-(e)$; then e, e^*, h^* are in reverse optical line and $S \in \mathcal{R}$ so the Signal Function Inequality (Th.39(ii)) implies

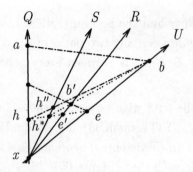

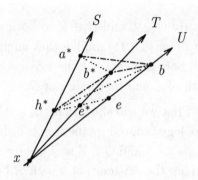

Figure 65   Figure 66

that $e^* \leq b^*$. The Signal Function Inequality (Th.39(i)), the Triangle Inequality (Th.40(i)) and the weak order-preserving property of signal functions imply that

$$f^+_{ST}(b^*) = a^* \quad \text{and} \quad f^-_{ST}(b^*) = h^* \ .$$

We now see that the configuration S, T, e^*, h^*, b^*, a^* has order and signal functions corresponding to the configuration Q, U, e, h, b, a of Case (α): it therefore implies a similar contradiction.

In summary, the supposition that $b > e$ implies contradictions for both Cases (α) and (β), so we conclude that $b = e$.

(b) In the general case, we consider any two paths Q, R and any event $e \in R$. The Second Existence Theorem (Th.14) implies the existence of a path U which meets the path R at the event e and the path Q at an event x before e. By (a) above and the definition of signal functions

$$h := f^-_{QU}(e) = f^-_{QR}(e) \quad \text{and} \quad e = f^+_{UQ}(h)$$

so, by the Signal Function Inequality (Theorem 39(ii)),

$$f^+_{RQ}(h) \geq f^+_{RQ} \circ f^-_{QR}(e) \geq e$$

and by the Triangle Inequality (Theorem 40(i))

$$f^+_{RQ}(h) \leq f^+_{RU} \circ f^+_{UQ}(h) = e \ ,$$

whence $f^+_{RQ}(h) = e$. But R was arbitrary and $e \in R$ was arbitrary, hence

$$f^-_{RQ} = (f^+_{QR})^{-1}$$

(c) The forward direction of time has been chosen arbitrarily, so the proof of the first equality of (i) also establishes the second equality. Hence both forward and reverse signal functions are one-to-one onto mappings (part (ii)). Statements (iii) and (iv) follow directly from the order-preserving property of signal functions (Theorem 38). To obtain the result (v) we observe that the Axiom of Continuity (Axiom C) implies the monotonic sequence property (namely that a bounded monotone sequence has a limit), and then the equality of (v) is an immediate consequence of (iv). *q.e.d.*

6.4 Optical lines in a collinear set

If three events a, b, c of Σ are signal–related such that

$$a \, \sigma \, b \quad \text{and} \quad b \, \sigma \, c \quad \text{and} \quad a \, \sigma \, c$$

we say (as before in the statement of the Second Optical Line Theorem (Th.41)), that the events are *in optical line* and we denote this by writing $|a, b, c\rangle$ where the notation indicates the "direction" of the (forward) signal relation. Thus we may think of "a signal which goes from a to b and then to c". Similarly, a set of events $\{Q_1^{(1)}, Q_2^{(2)}, \dots, Q_n^{(n)}\}$ of Σ is *in optical line* if and only if, for all a, b, c with $1 \le a \le b \le c \le n$

$$|Q_a^{(a)}, Q_b^{(b)}, Q_c^{(c)}\rangle$$

and we write

$$|Q_1^{(1)}, Q_2^{(2)}, \dots, Q_n^{(n)}, \dots\rangle \, .$$

(Physically, a set of events of $\mathcal{M}$ would be in optical line if they lay on a "light ray". The existence of such maximal sets of signal–related events is proved much later (see Section 10.2) within the present axiomatic system since, at this stage, we can not yet assert that $|a, b, c\rangle$ implies the collinearity of the events a, b, c. Thus it would be premature to speak of a "light ray". However we can establish useful results if we restrict our attention to collinear sets as we do in this chapter and the following chapters 7 and 8). Anticipating the results of the following theorem we will define an *optical line* to be a maximal subset of a collinear set such that any three events are in optical line.

In the next theorem we demonstrate the existence of *right (directed) optical lines* which will be referred to simply as *right optical lines*, and also *left optical lines*: the definitions are included in the statement of part (i) of the theorem.

In the proof of the next theorem, it will be convenient to refer to a collinear sub–spray by indicating the event x at which its corresponding paths meet: thus we shall use the notations $csp[x]$, $csp^+[x]$ and $CSP[x]$ when it is clear that the discussion refers to one particular collinear set.

Theorem 44 (Third Optical Line)
Let Σ be a collinear set, let Q be any path in Σ and let Q_y be any event of Q.

(i) *There are two distinct optical lines, each containing Q_y and exactly one event from each path in Σ (other than Q). The optical line which contains events in the order from left to right is called a right-directed optical line or simply a right optical line. The other optical line is called a left optical line. Along any optical line the relations "to the right of" and "to the left of" are transitive.*

(ii) *Each optical line is simply-ordered, has no first or last event, and is dense in itself.*

Proof (i) Let $Q_w \in Q$ be an event such that $Q_w < Q_y$. Then Σ contains a collinear subspray $csp^+[Q_w]$, whose paths meet at Q_w. The Signal Theorem (Th.43) and the Second Optical Line Theorem (Th.41) imply that there are two distinct optical lines in $csp^+[Q_w]$ each containing Q_y.

Now for any path $R \subset \Sigma$ which does not meet Q at Q_y we can take Q_v such that

$$Q_v < F_{RQ}^{-1} \circ F_{QR}^{-1}(Q_y)$$

and then both $F_{QR}^{-1}(Q_y)$ and $F_{RQ}(Q_y)$ are after Q_v so both events are in $csp^+[Q_v]$ (refer to part (i) of the proof of the Collinear Set Theorem (Th.36)) and hence each of these two events lies on an optical line through Q_y in $csp^+[Q_v]$. Now (as in part (i) of the proof of the Collinearity Theorem)

$$Q_u < Q_v \implies csp^+[Q_u] \supset csp^+[Q_v]$$

and

$$\Sigma = \bigcup_{Q_u \in Q} csp^+[Q_u]$$

hence each optical line through Q_y in $csp^+[Q_v]$ defines a unique optical line in Σ. Since the unreachable set $R(Q_y, \emptyset)$ is connected (Theorem 13), the Crossing Theorem (Th.37) implies that both $F_{QR}^{-1}(Q_y)$ and $F_{RQ}(Q_y)$ are on the same side of Q. If they are on the right side the optical line containing the events related by

$$Q_y \; \sigma \; F_{RQ}(Q_y)$$

is called a *right-directed optical line* or simply a *right optical line* and the optical line containing the events related by

$$F_{QR}^{-1}(Q_y) \ \sigma \ Q_y$$

is called a *left-directed optical line* or simply a *left optical line*; while if the events are both on the left side of Q the words "left" and "right" are interchanged in the above definition. The Crossing Theorem (Th.37(iii)) implies that the relations "to the right of" and "to the left of" are transitive along any optical line.

(ii) Now for each $Q_z \in Q$ with $Q_z < Q_y$, consider the restriction of each optical line to the events of $csp^+[Q_z]$. The Second Optical Line Theorem (Th.41) implies simple ordering, the Intermediate Path Theorem (Th.31) implies density and Theorem 33(iii) implies that there is no last event. Since the forward direction of time is chosen as a convention, it follows that there is also no first event. $\quad$ *q.e.d.*

6.5 Reflection in a path

Given a collinear set Σ and a path Q in Σ, we say that the events e_1 and e_2 are *reflections* of each other in Q if they are on opposite sides of Q in Σ and if there are two events $Q_a, Q_b \in Q$ which are signal–related to both e_1 and e_2; that is if $Q_a \ \sigma \ e_1 \ \sigma \ Q_b$ and $Q_a \ \sigma \ e_2 \ \sigma \ Q_b$. We say that two paths S, T are *reflections* of each other in Q if their sets of events are reflections in Q and we indicate this by writing $S = T_Q$ or $T = S_Q$.

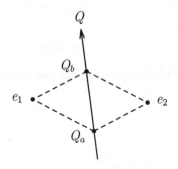

Figure 67 $\quad$ The events e_1 and e_2 are reflections of each other in the path Q: dashed lines indicate a signal relation between events

Theorem 45 (Reflection Mapping)

Given any collinear set Σ and any path Q in the collinear set, there is a unique bijective reflection mapping ψ of the collinear set onto itself such that:

(i) $T \subset \Sigma \Longrightarrow \psi(T) \subset \Sigma$

(ii) *for any paths T, U in Σ and any event $T_x \in T$,*

$$
\begin{aligned}
F_{UT}(T_x) &= U_y &&\Longrightarrow F_{\psi(U)\psi(T)}(\psi(T_x)) &&= \psi(U_y) \\
F_{TU} \circ F_{UT}(T_x) &= T_z &&\Longrightarrow F_{\psi(T)\psi(U)} \circ F_{\psi(U)\psi(T)}(\psi(T_x)) &&= \psi(T_z)
\end{aligned}
$$

(iii) *ψ is order–preserving*

(iv) *ψ maps parallel paths to parallel paths of the same type.*

Remarks 1. The mapping ψ is called a *reflection mapping.* It leaves all events of Q invariant.

2. This theorem has an obvious extension to an isotropy mapping between collinear sets Σ_1 and Σ_2 which have a common path Q.

Proof (see Figures 68 and 69). (i) We first show that there is a reflection mapping ψ. As in the proofs of previous theorems, for any event $a \in Q$, we can demonstrate the existence of paths T, R which meet Q at a and which lie on opposite sides of Q after the event a. If, for some event $d \in Q$ with $d > a$,

$$
F_{QT} \circ F_{TQ}(d) < F_{QR} \circ F_{RQ}(d) \tag{1}
$$

the Third Existence Theorem (Th.42) implies the existence of a path U between Q and R such that

$$
F_{QT} \circ F_{TQ}(d) = F_{QU} \circ F_{UQ}(d) =: e
$$

so

$$
Q(e, T, a, \emptyset) = Q(e, U, a, \emptyset)
$$

and by the Axiom of Isotropy (Axiom S) there is an isotropy mapping $\theta_a : \mathcal{E} \longrightarrow \mathcal{E}$ which induces a mapping of Σ to itself and a bijection from the set of paths of Σ to itself such that $U = \theta_a(T)$. (A similar argument applies if the inequality of (1) is reversed or replaced by an equality).

For any path V in Σ and any event $V_y \in V$, the Signal Theorem (Th.43) implies that there are unique events $Q_x, Q_z \in Q$ such that

$$
Q_x \; \sigma \; V_y \; \sigma \; Q_z \; . \tag{2}
$$

Since θ_a induces a bijective map from $\mathcal{P}$ to $\mathcal{P}$ and leaves the events of Q invariant (Axiom S(ii)), it follows that

$$
Q(V_y, \emptyset) = Q(\theta_a(V_y), \emptyset) \tag{3}
$$

whence

$$Q_x \circ \theta_a(V_y) \circ Q_z \, . \tag{4}$$

Equations (2) and (4) show that signal functions from Q to any path V are mapped onto signal functions from Q to $\theta_a(V)$.

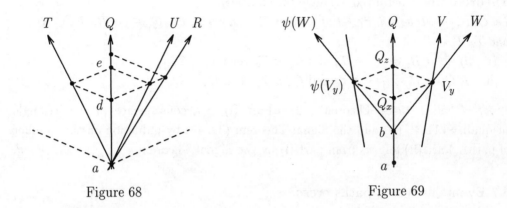

Figure 68   Figure 69

To show that the reflection mapping θ_a is independent of the event a, we consider, firstly, an event $V_y > a$. Then the Density Theorem (Th.17) implies the existence of an event $b \in Q$ such that $a < b < V_y$, so there is a path W which passes through both b and V_y. Now $V_y \in W$ implies that $\theta_a(V_y) \in \theta_a(W)$ and, by the Third Existence Theorem (Th.42), the two paths W and $\theta_a(W)$ are the only paths in Σ which pass through b and whose record functions relate Q_x to Q_z. Thus the images $\theta_a(W)$ and $\theta_a(V_y)$ are independent of the choice of the event a. The subset $\{V_y : V_y > b, V_y \in V\}$ is mapped by θ_a onto a subset of events of the path $\theta_a(V)$, so the Axiom of Uniqueness (Axiom I3) implies that the path $\theta_a(V)$ is also determined independently of the choice of the event a. Secondly, since any event of Σ can be specified by the intersection of two paths, the image of each event of Σ is specified independently of the choice of a by the intersection of the images of the two paths.

(ii) The special cases where $T = Q$ have been established by equations (2), (3) and (4) in the proof of (i). The first result is obtained from the special cases where $T = Q$ by a simple application of the Third Optical Line Theorem (Th.44). The second result is a corollary of the first.

(iii) This is a direct consequence of (i) and (ii).

(iv) The reflection mapping is bijective and order–preserving, so the defining properties for parallelism are preserved.   *q.e.d.*

6.6 Generalized triangle inequalities

The next theorem generalizes the previous Triangle Inequalities (Theorem 40) and is due to Walker (1948, Theorem 5.2) whose derivation was based on axioms equivalent to the Triangle Inequality and the Signal Theorem.

Theorem 46 Generalized Triangle Inequalities)
Let $Q, R, \cdots, T$ be a finite set of paths in Σ with events $Q_x, Q_y \in Q$; $R_x, R_y \in R$; $\cdots$ and $T_x, T_y \in T$.

   (i)   *If $Q_x \; \sigma \; R_y \; \sigma \; \cdots \; \sigma \; T_y$ and $Q_x \; \sigma \; T_x$ then $T_x \leq T_y$*

   (ii) *If $Q_x \; \sigma \; R_y \; \sigma \; \cdots \; \sigma \; T_y$ and $Q_y \; \sigma \; T_y$ then $Q_x \leq Q_y$*

Proof (Walker 1948, Theorem 5.2). Part (i) is a consequence of the Triangle Inequality (Th.40 (i)) and the Signal Theorem (Th.43) by induction on the number of paths. Part (ii) follows from part (i) by the Signal Theorem.   *q.e.d.*

6.7 Events at which paths cross

This theorem is based on a theorem of Walker (1948, Theorem 8.1). It is a stronger result than Walker's since it states that the paths not only meet but also cross. The method of proof is different.

Theorem 47 (Events at which paths cross)
Given distinct paths Q, T in Σ and an event $T_0 \in T \backslash Q$, we define (for all integers n),

$$T_n := (F_{TQ} \circ F_{QT})^n (T_0) .$$

If $\lim_{n \to \infty} T_n \in T$ we let $T_\infty := \lim_{n \to \infty} T_n$ and if $\lim_{n \to -\infty} T_n \in T$ we let $T_{-\infty} := \lim_{n \to -\infty} T_n$.

   (i)   *If T_∞ exists then Q crosses T at T_∞.*

   (ii) *If $T_{-\infty}$ exists then Q crosses T at $T_{-\infty}$.*

  (iii) *For all $T_x \in \bigcup_{n \to -\infty}^\infty |T_n T_{n+1}|$, the paths Q and T do not cross (and do not meet)*

(see Figure 70).

Proof Let $Q_n := F_{QT}(T_n)$ and define the function $q := F_{TQ} \circ F_{QT}$. Since the unreachable set $T(Q_n, \emptyset)$ is connected (Theorem 13), the Crossing Theorem (Th.37) implies that both T_{n-1} and T_n are on the same side of Q so that Q and T do not cross (and do not meet) in the interval $|T_{n-1} T_n|$: similar considerations apply for all n, which proves (iii).

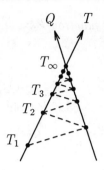

Figure 70   The set $\{T_n\}$ has a supremum $T_\infty \in T$ and the paths cross at T_∞

If T_∞ exists, the Signal Theorem (Th.43(v)) implies that

$$q(T_\infty) = q(\lim_{n \to \infty} T_n) = \lim_{n \to \infty} (q(T_n)) = \lim_{n \to \infty} (T_{n+1}) = T_\infty \;.$$

which, together with Axiom I5 shows that T and Q meet at T_∞. The Crossing Theorem 37 implies that they cross, which establishes (i): the proof of (ii) is similar.

$$q.e.d.$$

6.8 Modified signal and record functions

Walker (1948) discussed collinear sets using both "record functions" and "modified record functions". We also define "modified signal functions" as in the previous axiomatic system of Schutz (1973). Modified record and signal functions incorporate a "sense of direction" which simplifies subsequent proofs and leads to the eventual coordinatization of a collinear set.

Given two paths Q, R in a collinear set, the *record function* of the path R with respect to Q is defined to be the composition of signal functions $F_{QR} \circ F_{RQ}$. The *modified record function* $(F_{QR} \circ F_{RQ})^*$ is defined:

$$(F_{QR} \circ F_{RQ})^*(Q_x) := \begin{cases} F_{QR} \circ F_{RQ}(Q_x) & \text{, if } R \text{ is to the right of } Q_x \\ Q_x & \text{, if } R \text{ coincides with } Q \text{ at } Q_x \\ (F_{QR} \circ F_{RQ})^{-1}(Q_x) & \text{, if } R \text{ is to the left of } Q_x \;. \end{cases}$$

The modified record function indicates relative position, for

$$F_x \underset{<}{\overset{>}{=}} Q_y$$

depending on whether R is to the right of, coincident with, or to the left, of Q at Q_x.

We define the *modified signal function* F_{RQ}^+, which is related to right optical lines, in a similar manner as follows:

$$F_{RQ}^+(Q_x) := \begin{cases} F_{RQ}(Q_x) & \text{, if } F_{RQ}(Q_x) \text{ is to the right of } Q_x \\ Q_x & \text{, if } R \text{ coincides with } Q \text{ at } Q_x \\ F_{QR}^{-1}(Q_x) & \text{, if } F_{QR}^{-1}(Q_x) \text{ is to the left of } Q_x . \end{cases}$$

Similarly, we define the *modified signal function* F_{RQ}^-, which is related to left optical lines, as follows:

$$F_{RQ}^-(Q_x) := \begin{cases} F_{RQ}(Q_x) & \text{, if } F_{RQ}(Q_x) \text{ is to the left of } Q_x \\ Q_x & \text{, if } R \text{ coincides with } Q \text{ at } Q_x \\ F_{QR}^{-1}(Q_x) & \text{, if } F_{QR}^{-1}(Q_x) \text{ is to the right of } Q_x . \end{cases}$$

Theorem 48 *For paths* Q, R, S *in* Σ

(i) $\qquad (F_{RQ}^+)^{-1} = F_{QR}^+ \quad , \qquad (F_{RQ}^-)^{-1} = F_{QR}^- \, ,$

(ii) $\qquad F_{SR}^+ \circ F_{RQ}^+ = F_{SQ}^+ \quad , \qquad F_{SR}^- \circ F_{RQ}^- = F_{SQ}^- \quad and$

(iii) $\quad (F_{QR} \circ F_{RQ})^* \; = \; F_{QR}^- \circ F_{RQ}^+ \, .$

Proof The results (i) and (ii) are consequences of the previous definitions and the Third Optical Line Theorem (Th.44). To establish (iii), we consider separately the possibilities of R being to the right of Q, coincident with Q (which is not shown since it is trivial), and to the left of Q. We apply the previous definitions. Thus

$$(F_{QR} \circ F_{RQ}) = \begin{cases} F_{QR} \circ F_{RQ} \\ (F_{QR} \circ F_{RQ})^{-1} \end{cases}$$

$$= \begin{cases} F_{QR} \circ F_{RQ} \\ (F_{RQ})^{-1} \circ (F_{QR})^{-1} \end{cases}$$

$$= \begin{cases} F_{QR} \\ (F_{RQ})^{-1} \end{cases} \circ \begin{cases} F_{RQ} \\ (F_{QR})^{-1} \end{cases}$$

$$= F_{QR}^- \circ F_{RQ}^+ \, .$$

$q.e.d.$

Theorem 49 (Order of Events on an Optical Line)

Let Q, R, S be paths in Σ and let $Q_x \in Q$.

  (i) *The order of the events $Q_x, F^+_{RQ}(Q_x), F^+_{SQ}(Q_x)$ on the right optical line through Q_x is the same as the order of the events.*

$$Q_x \ , \ (F_{QR} \circ F_{RQ})^*(Q_x) \ , \ (F_{QS} \circ F_{SQ})^*(Q_x) \quad in \ Q \ .$$

  (ii) *If $(F_{QR} \circ F_{RQ})^*(Q_x) = (F_{QS} \circ F_{SQ})^*(Q_x)$, then R and S meet at $F^+_{RQ}(Q_x)$.*

Remark   This theorem is based on theorems of Walker (1948, Theorems 8.2 and 8.3).

Proof   Both results are immediate consequences of the definitions and the preceding theorem which implies, for (ii), that

$$(F_{SR} \circ F_{RS})^*(F^+_{SQ}(Q_x)) = F^+_{SQ}(Q_x) \ .$$

$$q.e.d.$$

6.9 Mid–way and reflected paths

In this section we will demonstrate the existence of a path "mid–way between" two paths of a collinear sub–SPRAY. Then we establish the existence of "reflections". This leads to the existence of sets of paths, indexed by dyadic numbers which correspond to the measure of "speed" or "rapidity of separation". The first theorem is a description of relative "closeness" expressed in terms of record functions.

Theorem 50 *Let Q, R, S, U be distinct paths in Σ which meet at an event x.*

  (i) *If for some $Q_0 > x$,*

$$F_{QR} \circ F_{RQ}(Q_0) < F_{QS} \circ F_{SQ}(Q_0)$$

then for all $Q_y > x$,

$$F_{QR} \circ F_{RQ}(Q_y) < F_{QS} \circ F_{SQ}(Q_y)$$

and

$$(F_{QR} \circ F_{RQ})^{-1}(Q_y) > (F_{QS} \circ F_{SQ})^{-1}(Q_y) \ .$$

  (ii) *If, additionally, $\langle Q, R, U \rangle$ and $\langle Q, S, U \rangle$, then $\langle Q, R, S, U \rangle$.*

Remark   There is also a result which corresponds to (i) for equality of record functions.

Proof Let us specify the right side of Q in Σ to be the side which contains S after x. The Reflection Mapping Theorem (Th.45) implies the existence of a reflected path $R^* := R_Q$. Since collinear sub–SPRAYS are simply ordered (Theorem 33), the previous theorem (Th.49) implies that this ordering corresponds to the order of the modified record functions, so that either $\langle R^*, Q, R, S \rangle$ or $\langle R, Q, R^*, S \rangle$. The Reflection Mapping Theorem implies that both R and R^* have the same record function with respect to Q so the previous theorem stated in terms of (ordinary, i.e. not modified) record functions, implies the stated ordering of record functions for all $Q_y > x$. The result for the inverse of record functions is a consequence of the strict order preserving property of record functions which is implied by the corresponding property for signal functions (Theorem 43(iii)). Part (ii) is a simple consequence of the preceding theorem (Th.49). *q.e.d.*

Let Q, S, U be paths which coincide at the event Q_c. If $\langle Q, S, U \rangle$ and if

$$F_{SQ} \circ F_{QS} = F_{SU} \circ F_{US},$$

(as in Figure 71) we say that S is *mid–way between* Q and U.

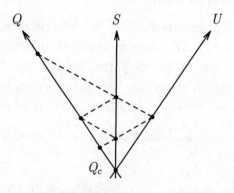

Figure 71

Theorem 51 (Existence of Mid–Way Path)
Let Q, U be distinct paths which coincide at the event Q_c. Then there is a path S mid–way between Q and U. That is

$$Q = U_S \quad and \quad U = Q_S$$

and furthermore

$$F_{QU} \circ F_{UQ} = (F_{QS} \circ F_{SQ})^2 .$$

106

Proof   Theorem 33 implies that there is a simply ordered set of paths

$$\mathcal{C} = \{R : \langle Q, R, U \rangle\} \cup \{Q, U\}$$

and as in the proof of the Third Existence Theorem (Th.42), the set $\mathcal{C}$ is complete with respect to the order topology as implied by the Continuity Theorem (Th.12). Let Q_x be an event after Q_c and let $Q_z := (F_{QU} \circ F_{UQ})(Q_x)$. By the Third Existence Theorem and Theorem 50 there is a path $S \in \mathcal{C}$ such that

$$F_{QS} \circ F_{SQ}(Q_x) = \sup\{F_{QR} \circ F_{RQ}(Q_x) : (F_{QR} \circ F_{RQ})^2(Q_x) \le Q_z, R \in \mathcal{C}\}$$
$$=: Q_1 . \tag{1}$$

Case (a) (See Figure 72). Suppose $(F_{QS} \circ F_{SQ})^2(Q_x) < Q_z$. Then the Density Theorem (Th.17) and the Third Existence Theorem imply the existence of an event $Q_3 \in Q$ and a path $V \in \mathcal{C}$ such that

$$Q_1 < (F_{QV} \circ F_{VQ})^{-1}(Q_z) = Q_3 < (F_{QS} \circ F_{SQ})^{-1}(Q_z),$$

whence the previous theorem (Th.50) implies that $\langle Q, S, V, U \rangle$. Again, by the Third Existence Theorem there is a path $T \in \mathcal{C}$ such that

$$F_{QT} \circ F_{TQ}(Q_x) = \min\{Q_3, F_{QV} \circ F_{VQ}(Q_x)\} > Q_1$$

so the previous theorem implies that

$$\langle Q, S, T, V, U \rangle \quad \text{or} \quad T = V$$

and hence that

$$(F_{QT} \circ F_{TQ})^2(Q_x) \le F_{QT} \circ F_{TQ}(Q_3)$$
$$\le F_{QV} \circ F_{VQ}(Q_3) = Q_z ,$$

which contradicts the definition of S in (1), and shows that the supposition is false.

Case (b) (See Figure 72). Suppose $(F_{QS} \circ F_{SQ})^2(Q_x) > Q_z$. Then there is an event $Q_3 \in Q$ and a path $V \in \mathcal{C}$ such that

$$Q_1 > (F_{QV} \circ F_{VQ})^{-1}(Q_z) = Q_3 > (F_{QS} \circ F_{SQ})^{-1}(Q_z) ,$$

which, by the previous theorem, means that $\langle Q, V, S, U \rangle$. Now by the Third Existence Theorem (Th.42) there is a path $T \in \mathcal{C}$ such that

$$F_{QT} \circ F_{TQ}(Q_x) = \max\{Q_3, F_{QV} \circ F_{VQ}(Q_x)\} < Q_1$$

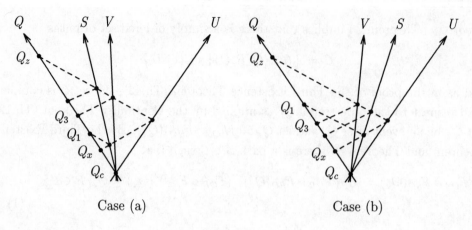

Case (a)   Case (b)

Figure 72

and $\langle Q, V, T, S \rangle$ or $V = T$. But now

$$(F_{QT} \circ F_{TQ})^2(Q_x) \geq F_{QT} \circ F_{TQ}(Q_3)$$
$$\geq F_{QV} \circ F_{VQ}(Q_3) = Q_z,$$

which contradicts the definition of S in (1) and shows that the supposition is false.

Having eliminated the previous two cases, we conclude that

$$(F_{QS} \circ F_{SQ})^2(Q_x) = Q_z = F_{QU} \circ F_{UQ}(Q_x)$$

Also $\langle Q, S, U \rangle$ so letting $S_x := F_{SQ}(Q_x)$ and $S_z := F_{QS}^{-1}(Q_z)$,

$$F_{SQ} \circ F_{QS}(S_x) = S_z = F_{SU} \circ F_{US}(S_x) \ .$$

By the Reflection Mapping Theorem (Th.45)

$$F_{SQ} \circ F_{QS} = F_{SU} \circ F_{US} \ ,$$

that is

$$Q = U_S \quad \text{and} \quad U = Q_S \ .$$

The Modified Signal and Record Function Theorem (Th.48) then implies that

$$F_{QU} \circ F_{UQ} = \left(F_{QS} \circ F_{SQ} \right)^2 \ .$$

q.e.d.

The Reflection Mapping Theorem (Th.45) implies the existence of reflected paths Q_U and U_Q. The next theorem establishes the existence of "multiple reflections".

Theorem 52 (Integer Indexed sub–SPRAY)

Let S^0, S^1 be distinct paths in Σ which coincide at the event S_c^0. Then there is a sub–SPRAY $\{S^n : n = 0, \pm 1, \pm 2, \ldots; \ S^n \in SPR[S_c^0]\}$ in Σ such that, for all integers m, n

$$\left(F_{mn} \circ F_{nm}\right)^* = \left(F_{m(m+1)} \circ F_{(m+1)m}\right)^{n-m},$$

where $F_{mn} := F_{S^m S^n}$.

Remarks

1. See Figure 73.
2. An immediate consequence of the previous theorem is that this result can be extended to dyadic numbers $m = M/2^P$ and $n = N/2^Q$ where M, N are integers and P, Q are positive integers.

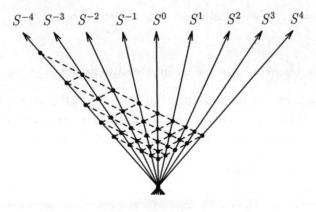

$$S^{-4} \quad S^{-3} \quad S^{-2} \quad S^{-1} \quad S^0 \quad S^1 \quad S^2 \quad S^3 \quad S^4$$

Figure 73

Proof (By induction). For $n > 1$, let S^{n+1} be the reflection of S^{n-1} in S^n. For $n \leq 0$, let S^{n-1} be the reflection of S^{n+1} in S^n.

Case (a) $m < n$. As an induction hypothesis, suppose that for all k, ℓ with $1 \le \ell - k \le n$,

$$F_{k\ell} \circ F_{\ell k} = \left(F_{k(k+1)} \circ F_{(k+1)k}\right)^{\ell-k}.$$

Then since $\langle S^k, S^{k+1}, S^{\ell+1} \rangle$, Theorems 48 and 49 imply that

$$
\begin{aligned}
F_{k(\ell+1)} \circ F_{(\ell+1)k} &= F_{k(k+1)} \circ F_{(k+1)(\ell+1)} \circ F_{(\ell+1)(k+1)} \circ F_{(k+1)k} \\
&= F_{k(k+1)} \circ \left(F_{(k+1)(k+2)} \circ F_{(k+2)(k+1)}\right)^{\ell-k} \circ F_{(k+1)k} \\
&= F_{k(k+1)} \circ \left(F_{(k+1)k} \circ F_{k(k+1)}\right)^{\ell-k} \circ F_{(k+1)k} \\
&= \left(F_{k(k+1)} \circ F_{(k+1)k}\right)^{\ell+1-k}.
\end{aligned}
$$

So if the induction hypothesis is true for all k, ℓ with $\ell - k \le n$, it is also true for all k, ℓ with $\ell - k \le n+1$. Since the induction hypothesis is (trivially) true for $n = 1$, the proof of this case is complete.

Case (b) $m > n$. In accordance with the definition of modified record functions, the proof is analogous to that of Case (a) but with inverse functions instead of functions.

Case (c) $m = n$. This trivial case completes the proof. $\qquad$ q.e.d.

6.10 Existence theorem for paths in a collinear set

We now establish two useful existence theorems for paths in a collinear set. The second theorem (Theorem 54) describes the existence of paths under the most general conditions and is expressed in terms of modified record functions.

Theorem 53 (Fourth Existence)
For any collinear set Σ, any path Q in Σ, and events $Q_c, Q_x, Q_y \in Q$ with $Q_c < Q_x < Q_y$ there are unique paths W, U in Σ with W on the left side of Q after Q_c and with U on the right side of Q after Q_c such that

$$F_{QW} \circ F_{WQ}(Q_x) = F_{QU} \circ F_{UQ}(Q_x) = Q_y.$$

Proof By the definition of a collinear set there is some path other than Q in Σ, so Theorem 16 and the Second Existence Theorem (Th.14) imply the existence of a path R which meets Q at Q_c and is to the right of Q after Q_c. We now let $T^0 := Q$, $T^1 := R$, $T_1^0 := Q_x$: then the previous theorem (Th.52) and the Crossing Theorem (Th.37) imply that there is some positive integer n such that

$$Q_c < Q_x = T_1^0 < Q_y \leq T_n^0$$

where T_n^0 is defined as in Theorem 47.

We now let $V := T^n$ (defined analogously to S^n of the previous theorem) and then the conditions of the Third Existence Theorem (Th.42) are satisfied so there is a path U satisfying the stated conditions: the Reflection Mapping Theorem (Th.45) implies the existence of a path $W := U_Q$.   q.e.d.

Theorem 54 (Existence of Paths in a Collinear Set)
Given a collinear set Σ, a path Q in Σ and any four events $Q_w, Q_x, Q_y, Q_z \in Q$ with $Q_w < Q_y$ and $Q_x < Q_z$, there is a unique path S in Σ such that

$$(F_{QS} \circ F_{SQ})^*(Q_w) = Q_x \quad and \quad (F_{QS} \circ F_{SQ})^*(Q_y) = Q_z .$$

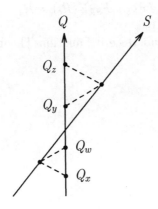

Figure 74   In this illustration $Q_w > Q_x$ and $Q_y < Q_z$

Remark   See Figure 74.

Proof   (see Figure 75). Take an event $Q_a \in Q$ with $Q_a < \min\{Q_w, Q_x, Q_y, Q_x\}$. The previous theorem (Th.53) implies the existence of a path R in Σ which passes through Q_a such that

$$(F_{QR} \circ F_{RQ})^*(Q_w) = Q_x .$$

111

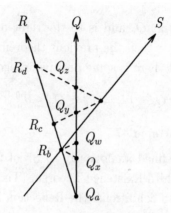

Figure 75

Let $R_b := F_{RQ}^+(Q_w)$, let $R_c := F_{RQ}^+(Q_y)$ and let $R_d := F_{RQ}^-(Q_z)$. Then the given order conditions and the Modified Signal and Record Function Theorem (Th.48) imply that $R_b < R_c$ and $R_b < R_d$, so the previous theorem implies the existence of a path S in Σ which passes through R_b such that

$$(F_{RS} \circ F_{SR})^*(R_c) = R_d$$

and so the Modified Signal and Record Function Theorem implies that S satisfies the stated conditions. *q.e.d.*

7.   Theory of parallels

In previous chapters we have demonstrated the existence of collinear sets of paths, and we have seen that collinear sets of paths have some properties analogous to coplanar sets of lines in the theory of absolute geometry. In the present chapter we shall see that the same properties of parallelism apply as in absolute geometry. Whereas a "parallel postulate" is required to distinguish between the Euclidean and Bolyai–Lobachevskian geometries, no special "parallel postulate" is required in the present treatment. However, until we prove the theorem which I take the liberty of naming the "Euclidean" Parallel Theorem (Theorem 68), we must consider the possibility of there being two different types of parallels.

Once we have shown that there is only one type of parallel, it will follow that each path has a "natural time-scale" which is determined to within an arbitrary increasing linear transformation. Then it is not difficult to show that modified signal functions are linear, and the ensueing discussion of one-dimensional kinematics is taken up in the next chapter.

In most of the subsequent proofs, questions of collinearity are trivial due to the results of preceding theorems.

7.1 Divergent and convergent parallels

The existence of parallel paths has already been established in Theorem 22. Now that there is a relation of temporal order on a collinear set Σ, it is convenient to refer to the two possible types of parallel paths as divergent parallels and convergent parallels and so we make the following definitions:

Given distinct paths Q, T in Σ and an event $U_y \in \Sigma$, we say that Q is a *divergent parallel from* T through the event U_y (Figure 76) and we write $Q \vee (T, U_y)$ if:

  (i) Q passes through the event U_y ,

  (ii) Q does not meet T at any event, and

  (iii) for each path $R \in SPR[U_y]$ such that $\langle R, Q, T \rangle$ after U_y the path R meets T at some event before U_y.

Sometimes we merely say that Q *diverges from* T through U_y .

Similarly, we say that Q is a *convergent parallel to* T through the event U_y (Figure 76) and we write $Q \wedge (T, U_y)$ if:

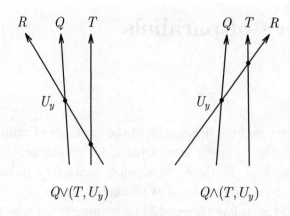

$$Q \vee (T, U_y) \qquad\qquad Q \wedge (T, U_y)$$

Figure 76

(i) Q passes through the event U_y ,

(ii) Q does not meet T at any event, and

(iii) for each path $R \in SPR[U_y]$ such that $\langle R, Q, T \rangle$ before U_y, the path R meets T at some event after U_y.

Sometimes we merely say that Q *converges to* T through U_y. In the remainder of this section we will often use the symbol $\|$ to represent either $\vee$ or $\wedge$, where it is implied that the substitution is consistent in any statement or proof. We now show that parallelism is a relation between paths by proving the following:

Theorem 55 (Transmissibility of Parallelism)
If $Q \| (T, U_y)$ and $Q_c \in Q$, then $Q \| (T, Q_c)$. That is, Q is parallel to T and we write $Q \vee T$ or $Q \wedge T$, as the case may be; or simply $Q \| T$ with the above convention.

Proof (a) *Transmissibility of Divergent Parallelism.* Suppose the contrary; that is, suppose there is a path R which meets Q at Q_c such that $\langle Q, R, T \rangle$ before Q_c . Now take an event $R_b \in R$ with $R_b > Q_c$ if $Q_c > U_y$ (Figure 77a) or with $R_b < Q_c$ if $Q_c < U_y$ (Figure 77b). By the Collinear Set Theorem (Th.36) there is a path S which meets Q at U_y and R at R_b . By the Crossing Theorem (Th.37), $\langle Q, S, R, T \rangle$ before $\min\{U_y, R_b\}$ so S does not meet T at any event before U_y ; but this contradicts the third requirement in the definition of Q.

(b) *Transmissibility of Convergent Parallelism.* The proof is similar to the proof of (a) with the expressions "before" and "min" changed to "after" and "max", respectively.   *q.e.d.*

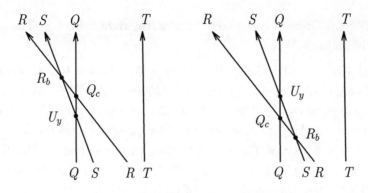

Figure 77a   Figure 77b

If we want to specify that Q diverges from T, or that Q converges to T, or simply that Q is parallel to T, we use the more concise expressions $Q \vee T, Q \wedge T$, and $Q \parallel T$ respectively. At this stage we have not shown that the relations of parallelism are symmetric or transitive. For convenience *we define both relations of parallelism to be reflexive*, in both the divergent and the convergent sense.

We now extend the definition of mid–way path so as to apply to paths, such as parallel paths, which need not meet at any event. Thus if Q, S, U are paths such that

$$\langle Q, S, U \rangle \quad \text{and} \quad F_{SQ} \circ F_{QS} = F_{SU} \circ F_{US} ,$$

we say that S is *mid–way between* Q and U: the previous definition of reflected paths applies here so $Q = U_S$ and $U = Q_S$.

We will now extend the result of Theorem 22 to demonstrate the existence of parallel paths of both types and their reflections.

Theorem 56 (Existence of Parallels and their Reflections)

Let S be a path in Σ.

(i) *For any event $V_0 \in \Sigma$ there are paths Q, U in Σ such that*

$$U \parallel (S, V_0) \quad and \quad Q = U_S .$$

(ii) *For any events $S_x, S_y \in S$ there is a path U in Σ with $U \parallel S$ such that*

$$(F_{SU} \circ F_{US})^*(S_x) = S_y .$$

115

Proof (i) We consider the case where V_0 is on the right side of S. The proof for the other case is similar.

Case (a) *Divergent Parallels*. We will specify events a, b, c, d to correspond to those of Theorem 22. Let $d = V_0$, then the Collinear Set Theorem (Th.36) implies the existence of events $b, c \in S$ and paths db, dc such that $c < d < b$. By Theorem 16, there is an event a and a path ca which (together with the events b, c, d) form the same configuration a, d, b, c as in Theorem 22. This theorem then implies the existence of a divergent parallel U from S through $d(= V_0)$, and the Reflection Mapping Theorem (Th.45) implies the existence of its reflection $Q = U_S$.

Case (b) *Convergent Parallels*. A similar proof applies.

(ii) By the Second Existence Theorem (Th.14) there is an event $S_w < \min\{S_x, S_y\}$ and Theorem 54 implies the existence of a path V which passes through S_w such that

$$(F_{SV} \circ F_{VS})^*(S_x) = S_y \; .$$

Now let $V_0 := F_{VS}^+(S_x)$; then part (i) above implies the existence of the parallel path U. $\hspace{6cm}$ *q.e.d.*

7.2 The parallel relations are equivalence relations

We now show that both relations of parallelism are equivalence relations. It will then follow that in any collinear set of paths, there are equivalence classes of parallel paths of both the divergent and the convergent type.

Theorem 57 (Symmetry of Parallelism)
Let Q, S be paths in Σ. If $Q \parallel S$ then $S \parallel Q$.

Proof We have already specified that the relation(s) of parallelism are reflexive (in the remarks following Theorem 55) so we consider distinct paths Q and S.

Case (a) *Divergent Parallels* (Figure 78). We suppose the contrary, namely that S does not diverge from Q; that is, we suppose that for some $S_a \in S$, there is a path U such that

$$U \vee (Q, S_a) \quad \text{and} \quad U \neq S \; . \tag{1}$$

By Theorem 51 there is a path T mid–way between S and U and a reflection mapping ϕ with invariant path T such that

$$\phi(T) = T \; , \quad \phi(S) = U \; , \quad \phi(U) = S \; . \tag{2}$$

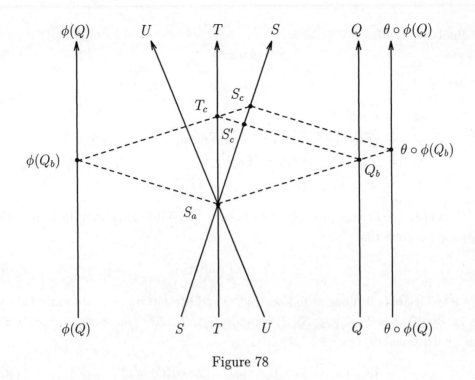

Figure 78

Now $\langle S, T, U, Q \rangle$ before S_a and $Q \vee S$ so

$$Q \vee S, \quad Q \vee T \quad \text{and} \quad Q \vee U . \tag{3}$$

Now by (3) and the Reflection Mapping Theorem (Th.45),

$$\phi(Q) \vee \phi(S), \quad \phi(Q) \vee \phi(T) \quad \text{and} \quad \phi(Q) \vee \phi(U) \tag{4}$$

so (2) implies that

$$\phi(Q) \vee U, \quad \phi(Q) \vee T \quad \text{and} \quad \phi(Q) \vee S . \tag{5}$$

Similarly (1) and (2) imply that

$$S \vee \phi(Q) . \tag{6}$$

Now consider a reflection mapping θ with invariant path S. Then the Reflection Mapping Theorem and (5) imply that

$$(\theta \circ \phi)(Q) \vee S \tag{7}$$

and similarly, by (6),

$$S \vee (\theta \circ \phi)Q \ . \tag{8}$$

Now let

$$
\begin{aligned}
Q_b &:= F_{QS}(S_a) \ , \\
S_c &:= \left(F_{S\phi(Q)} \circ F_{\phi(Q)S}\right)(S_a) = \left(F_{S(\theta\circ\phi)(Q)} \circ F_{(\theta\circ\phi)(Q)S}\right)(S_a) \\
S_c' &:= \left(F_{SQ} \circ F_{QS}\right)(S_a) = F_{SQ}(Q_b) \\
T_c &:= \left(F_{TQ} \circ F_{QT}\right)(S_a) = \left(F_{T\phi(Q)} \circ F_{\phi(Q)T}\right)(S_a) \ .
\end{aligned}
$$

By (1) and (2), $\langle U, T, S, Q \rangle$ after S_a and since $U = \phi(S)$ it follows from the Reflection Mapping Theorem that

$$\langle \ \phi(Q), U, T, S, Q \ \rangle \text{ after } S_a \ . \tag{9}$$

Since θ and ϕ are reflections, it follows that Q and $(\theta \circ \phi)(Q)$ are on the same side of S in Σ. By (9), $S \neq T$, so $|Q_b, \ S_c', \ T_c \ \rangle$ and $|\phi(Q_b), \ T_c, \ S_c \ \rangle$ and therefore $S_c' \sigma T_c \sigma S_c$. Thus by Theorem 49,

$$|S_a, \ Q_b, \ (\theta \circ \phi)Q_b \ \rangle \quad \text{and} \quad (\theta \circ \phi)(Q) \neq Q \ . \tag{10}$$

Since both Q and $(\theta \circ \phi)(Q)$ diverge from S, Theorem 55 implies that there is no event at which Q and $(\theta \circ \phi)(Q)$ can meet. Thus by (9), (10), the Crossing Theorem (Th.37) and Theorem 49,

$$\langle \phi(Q), S, T, U, Q, (\theta \circ \phi)(Q) \rangle \text{ before } S_a \ .$$

This is a contradiction of (8) since, by the supposition (1), $S \neq U$. This completes the proof for the case of divergent parallels.

Case (b) *Convergent Parallels.* A similar proof applies. $\hfill$ *q.e.d.*

Theorem 58 (Transitivity of Parallelism)
Let Q, R, S be paths in Σ. If $Q \parallel R$ and $R \parallel S$, then $Q \parallel S$.

Proof This proof is analogous to the proof of the corresponding theorem of absolute geometry.

Case (i) *Divergent Parallels.* The result is trivial unless the three paths are distinct, which is assumed from now on. We define the right side of Q to be the side which contains R. Now $Q \neq S$, so there is no event at which Q and S can meet.

Case (i)(a) $\langle Q, R, S \rangle$ (Figure 79a). Take any event $Q_c \in Q$, and any path $T \in SPR[Q_c]$ such that T is on the right side of Q before Q_c. Since $Q \vee R$, the path

T meets (and crosses) R at some event R_b before Q_c. Similarly, since $R \vee S$ and T is on the right side of R before R_b, the path T meets S at some event S_a before R_b. Since Q and S do not meet at any event, we conclude that $Q \vee S$.

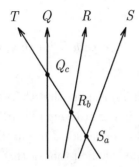

Figure 79a

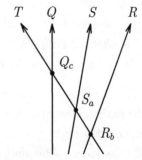

Figure 79b

Case (i)(b) $\langle Q, S, R \rangle$ (Figure 79b). Take any event $Q_c \in Q$ and any path $T \in SPR[Q_c]$ such that T meets R at some event R_a before Q_c. By the Crossing Theorem (Th.37), the path T crosses S at some event S_b between R_a and Q_c. Since Q and S do not meet at any event we conclude that $Q \vee S$.

Case (i)(c) $\langle S, Q, R \rangle$. The previous theorem implies that $S \vee R$ and $R \vee Q$ so interchanging the symbols "Q" and "S" in Case (i)(b), we find that $S \vee Q$. Again by the previous theorem (Th.57), $Q \vee S$. This completes the proof for divergent parallels.

Case (ii) *Convergent Parallels*. A similar proof applies with the word "before" and the symbol $\vee$ replaced by "after" and $\wedge$, respectively. *q.e.d.*

By definition the relations of parallelism are reflexive. The preceding two theorems show that they are symmetric and transitive. Thus they are equivalence relations on a collinear set. Since distinct parallel paths (of the same type) do not meet or cross, they can be ordered from left to right.

7.3 Coordinate systems on a collinear set

In this section we show that the set of events (of each path) is order–isomorphic to the real numbers and that the events of a collinear set may be coordinatized with respect to equivalence classes of parallels of either type.

Before discussing classes of parallel paths we first consider those subclasses of parallels which can be "indexed" by dyadic numbers (recall that a *dyadic number* is a number of the form $n/2^m$, where n is any integer and m is any non-negative integer).

Theorem 59 (Existence of Mid–Way Parallel)
Let Q, U be distinct paths in Σ with $Q \parallel U$. There exists a path S which is mid–way between Q and U and parallel to both.

Proof The Existence Theorem for Parallel Paths (Th.56) implies that there is a set C of parallel paths between and including Q and U. The linear ordering (Th.10) and the Completeness Property (Th.12) for events on Q apply to the set C of parallel paths due to the Existence Theorem (Th.56). The remainder of the proof of this theorem is essentially the same as that for the previous (Mid–Way Path) Theorem (Th.51), with Theorem 56 taking the place of Theorem 42. *q.e.d.*

Theorem 60 *Let S^0, S^1 be distinct paths with $S^0 \parallel S^1$.*
There is a collinear class of parallel paths

$$\{S^d : d \text{ is a dyadic number}\}$$

such that, for any integers m and n,

$$\left(F_{S^{dm}S^{dn}} \circ F_{S^{dn}S^{dm}}\right)^* = \left(F_{S^{dm}S^{d(m+1)}} \circ F_{S^{d(m+1)}S^{dm}}\right)^{n-m}$$

and for any dyadic numbers a, b, c,

$$a < b < c \iff \langle S^a, S^b, S^c \rangle .$$

Proof By the above theorem there is a path $S^{1/2}$ mid–way between S^0 and S^1 and by induction there is a path

$$S^{2^{-(m+1)}} \quad \text{mid–way between} \quad S^0 \quad \text{and} \quad S^{2^{-m}}. \tag{1}$$

We define the right side of S^0 to be the side which contains S^1. In the remainder of this proof we shall use the notation:

$$\rho\langle a; b \rangle := \left(F_{S^a S^b} \circ F_{S^b S^a}\right)^* ,$$

(where the superscripts a and b are not necessarily numbers). As in the Integer Indexed sub–SPRAY Theorem (Th.52), for each positive integer p we define a set of parallel paths (indexed by the integers),

$$\{S^{n/2^p} : n = 0, \pm 1, \pm 2, \dots\}$$

which has the property:

$$\rho \langle m/2^p; n/2^p \rangle = \rho^{n-m} \langle m/2^p; (m+1)/2^p \rangle \tag{2}$$

Now by (1),

$$\rho^2 \langle 0; 1/2^{(p+1)} \rangle = \rho \langle 0; 1/2^p \rangle \,,$$

and so by induction,

$$\rho^{2^q} \langle 0; 1/2^{(p+q)} \rangle = \rho \langle 0; 1/2^p \rangle \,.$$

Therefore, for any integer n,

$$\rho^{n.2^q} \langle 0; 1/2^{(p+q)} \rangle = \rho^n \langle 0; 1/2^p \rangle \,,$$

and by (2),

$$\rho \langle 0; n.2^q/2^{(p+q)} \rangle = \rho \langle 0; n/2^p \rangle \,.$$

Now by Theorem 49, for all integers n and for any non-negative integers p and q,

$$S^{n.2^q/2^{(p+q)}} = S^{n/2^p} \,,$$

so for each dyadic number there is a unique path and equation (2) is equivalent to the equation which was to be proved; the ordering property is trivial.   *q.e.d.*

A subclass of (convergent or divergent) parallels indexed by dyadic numbers will be called a *dyadic class of parallels*. We can define a *dyadic class of events* of the path S^0, by taking any particular event of S^0 and giving it the index S_0^0, and then letting

$$S_{2p}^0 := \left(F_{S^0 S^p} \circ F_{S^p S^0} \right)^* \left(S_0^0 \right) \,,$$

for each dyadic number p. If it is clear from the context that we are referring to a particular class of parallels and events, we shall simply call them *dyadic parallels* and *dyadic events*, respectively.

A further consequence of the preceding theorem is that the dyadic subscripts (of the subset of dyadic events of the path S^0) are ordered in accordance with the ordering of the events they represent; that is, for any dyadic numbers a and b,

$$a < b \Longrightarrow S_a^0 < S_b^0 \,.$$

Theorem 61 *Any dyadic class of events of any path is a countable dense subset of (the set of events of) the path.*

Proof Let S^0 be a given path with a given subset of dyadic events. Consider the sequence of events

$$\left(S^0_{(2-1/2^n)}: \ n = 0, 1, \dots \right).$$

This sequence is bounded and strictly increasing and therefore has a supremum

$$S^0_\omega := \sup\left\{ S^0_{(2-1/2^n)}: \ n = 0, 1, 2, \ \dots \right\}. \tag{1}$$

We will first show that $S^0_\omega = S^0_2$. Let $S^1_{-1} := F^{-1}_{S^0 S^1}(S^0_0)$ and let $S^1_1 := F_{S^1 S^0}(S^0_0)$. By the previous theorem

$$\left(F_{S^1 S(1-1/2^n)} \circ F_{S(1-1/2^n) S^1}\right)^{2^n} (S^1_{-1}) = S^1_1. \tag{2}$$

By Theorem 56, there is a parallel path Q (convergent or divergent as the case may be) such that

$$\left(F_{S^0 Q} \circ F_{Q S^0}\right)(S^0_0) = S^0_\omega \tag{3}$$

and since $S^0_\omega = \sup\{S^0_{2-1/2^n}\} \le S^0_2$, Theorem 49 implies that for all positive integers n,

$$\langle S^0, S^{1-1/2^n}, Q, S^1 \rangle \qquad \text{or} \qquad Q = S^1$$

so for all positive integers n,

$$\left(F_{S^1 Q} \circ F_{Q S^1}\right)^{2^n} (S^1_{-1}) \le S^1_1,$$

whence

$$\sup_n \left(F_{S^1 Q} \circ F_{Q S^1}\right)^{2n} (S^1_{-1}) \le S^1_1,$$

and so by Theorem 47, the parallel paths Q and S^1 meet at the event

$$\sup_n \left\{\left(F_{S^1 Q} \circ F_{Q S^1}\right)^{2^n} (S^1_{-1})\right\}.$$

Therefore $Q = S^1$ and so by (3),

$$S^0_\omega = S^0_2. \tag{4}$$

The remaining part of this proof is based on the proof of a theorem of Walker (1948, Theorem 13.1, p.330). Given any two events $S^0_x, S^0_z \in S^0$ with $S^0_x < S^0_z$, we will find a dyadic event S^0_y such that $S^0_x < S^0_y < S^0_z$. The Density Theorem (Th.17) and Theorem 56 imply that there is a parallel U (divergent or convergent as the case may be) to the right of S^0 such that

$$S^0_x < u^{-1}(S^0_z) < S^0_z \qquad \text{where} \quad u := (F_{S^0 U} \circ F_{U S^0}) . \tag{5}$$

By Theorem 55, U can not meet S^0 at S^0_2, so by equation (1) there is some positive integer m such that

$$S^0_{(2-2^{-m+2})} > u^{-1}(S^0_2) . \tag{6}$$

We define the set of paths

$$\left\{ R^{(n)} : R^{(n)} = S^{2^{-n}}; \ n = 1, 2, 3, \ldots \right\}$$

and so

$$r_n^{-1}(S^0_2) = S^0_{(2-2^{-n+1})} > S^0_{(2-2^{-n+2})} \qquad \text{where} \quad r_n := F_{S^0 R^{(n)}} \circ F_{R^{(n)} S^0} . \tag{7}$$

By (6) and (7),

$$r_m^{-1}(S^0_2) > u^{-1}(S^0_2)$$

and since parallels can not cross (by Theorem 55)

$$r_m^{-1} > u^{-1} .$$

Thus from (5),

$$S^0_x < r_m^{-1}(S^0_z) < S^0_z . \tag{8}$$

Since $R^{(m)}$ does not meet S^0 at any event, Theorem 47 implies that the increasing and decreasing sequences

$$\left(r_m^n(S^0_2) : \ n = 0, 1, 2, \ldots \right) \quad \text{and} \quad \left(r_m^{-n}(S^0_2) : \ n = 0, 1, 2, \ldots \right)$$

are unbounded, so (the set of events of) S^0 is covered by the set of semi-closed intervals

$$\left\{ (r_m^n(S^0_2), r_m^{n+1}(S^0_2)] : \ n = 0, \pm 1, \pm 2, \ldots \right\} .$$

Therefore there is some integer p such that

$$r_m^p(S^0_2) < S^0_z \le r_m^{p+1}(S^0_2) . \tag{9}$$

Now from (8) and (9),

$$S^0_x < r_m^{-1}(S^0_z) \le r_m^{-1} \circ r_m^{p+1}(S^0_2) = r_m^p(S^0_2) < S^0_z$$

and by definition, $r_m^p(S^0_2)$ is a dyadic event. $\qquad q.e.d.$

Theorem 62 *The set of events of each path is order–isomorphic to the set of real numbers.*

Proof By the Prolongation Theorem (Th.6) paths do not have first or last events. The preceding theorem (Th.61) implies that each path has a countable dense subset of events. Together with the result of the Continuity Theorem (Th.12), these conditions are necessary and sufficient (Sierpinski 1965, XI, §10, Theorem 1) for a linearly ordered set to be order–isomorphic to the reals. $\qquad\qquad$ *q.e.d.*

In the next theorem we define *indexed classes of parallels* (of divergent or convergent type) with (corresponding) *divergent* or *convergent time scales*. This then leads to the definition of *position–time coordinate systems* of divergent or convergent type.

Theorem 63 (Indexed Class of Parallels)
A class of parallels in Σ and the events belonging to them can be indexed by the real numbers such that, for any real numbers a, b, c

(i) $$F^+_{cb}(S^b_a) = S^b_{a-b+c} \quad and \quad F^-_{cb}(S^b_a) = S^c_{a+b-c} \ ,$$

whence

(ii) $$\left(F_{bc} \ \circ \ F_{cb}\right)^* \left(S^b_a\right) \ = \ S^b_{a-2b+2c}$$

where we have introduced the notation: $F_{ab} := F_{S^a S^b}$ (see Figure 80).

Furthermore, for any real numbers x and t, there is a parallel path S^x and an event $S^x_t \in S^x$.

(iii) *An event S^x_t can be specified by two events S^0_{t-x}, S^0_{t+x} on the path S^0 and a modified record function :*

$$\left(F_{0x} \ \circ \ F_{x0}\right)^* \left(S^0_{t-x}\right) \ = \ S^0_{t+x} \ .$$

Remark A class of parallels, indexed in this way, is called an *indexed class of parallels*. The subscript indices of any parallel are said to constitute a *divergent or convergent time scale*, according as to whether the class of parallels is a divergent or convergent class. For an event S^x_t the superscript–subscript pair $(x; t)$ will be called the *position–time coordinates* of the event S^x_t. The set of all events in Σ indexed by the corresponding ordered pairs $\{S^x_t : x, t \in \mathbb{R}\}$ is called the *coordinate system* $\{S^x_t\}$ defined on the indexed class of parallels $\{S^x\}$. The event $(0; 0)$ is called the *origin in*

124

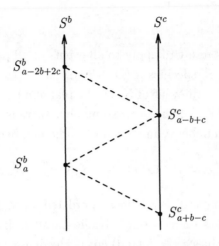

S^b S^c

$S^b_{a-2b+2c}$

S^c_{a-b+c}

S^b_a

S^c_{a+b-c}

Figure 80

position–time of the coordinate system and the set of events $\{(0;t) : t \in \mathbb{R}\}$ is called the *origin in position* of the coordinate system. A coordinate system is said to be *divergent* or *convergent* according as to whether the class of parallels is divergent or convergent.

Proof (i) We have already indexed the dyadic events of the path S^0 by letting

$$S^0_{2\gamma} := (F_{0\gamma} \circ F_{\gamma 0})^*(S^0_0) . \tag{1}$$

We will first show that, for any dyadic numbers α and β,

$$(F_{0\beta} \circ F_{\beta 0})^*(S^0_{2\alpha}) = (S^0_{2\alpha+2\beta}) . \tag{2}$$

Let d by a dyadic number (with $0 \le d \le 1$) such that there are integers m and n for which

$$\alpha = dm \quad \text{and} \quad \beta = dn .$$

Thus, by the definition (1) and Theorem 60,

$$
\begin{aligned}
(F_{0\beta} \circ F_{\beta 0})^*(S^0_{2\alpha}) &= (F_{0\ dn} \circ F_{dn\ 0})^* \circ (F_{0\ dm} \circ F_{dm\ 0})^*(S^0_0) \\
&= (F_{0d} \circ F_{d0})^n \circ (F_{0d} \circ F_{d0})^m (S^0_0) \\
&= (F_{0d} \circ F_{d0})^{m+n} (S^0_0) \\
&= (F_{0\ d(m+n)} \circ F_{d(m+n)\ 0})^*(S^0_0) \\
&= S^0_{2d(m+n)} \\
&= S^0_{2\alpha+\beta} ,
\end{aligned}
$$

which establishes equation (2).

We now extend this result to apply to all events of all parallels. By the previous theorem, any event of S^0 specifies a (Dedekind) cut in the dense subset of dyadic events of S^0. We index each event of S^0 by the real number so obtained; also by the Continuity Theorem (Th.12), for each real number there is a corresponding event of S^0; thus for each real number a and for any dyadic numbers α and $\overline{\alpha}$,

$$\alpha < a < \overline{\alpha} \quad \Longleftrightarrow \quad S^0_\alpha < S^0_a < S^0_{\overline{\alpha}} . \tag{3}$$

Futhermore, any class of parallels is a linearly ordered set so, by Theorem 56, between any two distinct parallels there is some dyadic parallel. In much the same way as above, any parallel specifies a (Dedekind) cut in the dense subset of dyadic parallels. We index each parallel by the real number so obtained; also by Theorem 56, for each real number b there is a parallel S^b such that

$$(F_{0b} \circ F_{b0})^*(S^0_0) = S^0_{2b} ;$$

thus, as above, for each real number b, and for all dyadic numbers β and $\overline{\beta}$,

$$\beta < b < \overline{\beta} \quad \Longleftrightarrow \quad (F_{0\beta} \circ F_{\beta 0})^* < (F_{0b} \circ F_{b0})^* < (F_{0\overline{\beta}} \circ F_{\overline{\beta} 0})^* . \tag{4}$$

Thus, by equations (2) and (3), for any given real number a and for all dyadic numbers α and $\overline{\alpha}$ such that $\alpha < a < \overline{\alpha}$,

$$S_{2\alpha + 2\beta} < (F_{0\beta} \circ F_{\beta 0})^*(S^0_{2a}) < S^0_{2\overline{\alpha} + 2\beta} ,$$

whence

$$(F_{0\beta} \circ F_{\beta 0})^*(S^0_{2a}) = S^0_{2a + 2\beta} .$$

Similarly, by the above equation and (4), for any real number b and for all dyadic numbers β and $\overline{\beta}$ such that $\beta < b < \overline{\beta}$,

$$S_{2a + 2\beta} < (F_{0b} \circ F_{b0})^*(S^0_{2a}) < S_{2a + 2\overline{\beta}} ,$$

whence

$$(F_{0b} \circ F_{b0})^*(S^0_{2a}) = S^0_{2a + 2b} . \tag{5}$$

We can now index the events of each path by defining, for all real numbers a and c,

$$S^c_a := F^+_{c0}(S^0_{a-c}) . \tag{6}$$

Now with the events of the parallels indexed in this way, equation (5) and Theorem 48 imply that

$$S^0_{a+c} = F^-_{0c}(S^c_a) \ . \tag{7}$$

Furthermore, by the same theorem,

$$F^+_{cb} = F^+_{c0} \circ (F^+_{b0})^{-1} \quad \text{and} \quad F^-_{cb} = F^-_{c0} \circ (F^-_{b0})^{-1} \ ,$$

so by (6) and (7),

$$F^+_{cb}(S^b_a) = S^c_{a-b+c} \quad \text{and} \quad F^-_{cb}(S^b_a) = S^c_{a+b-c} \tag{8}$$

whence, again by the same theorem

$$(F_{bc} \circ F_{cb})^*(S^b_a) = S^c_{a-2b+2c} \ . \tag{9}$$

This, together with equations (8), establishes (i) and (ii). Part (iii) is a special case of (ii). $\qquad q.e.d.$

The next theorem is analogous to a proposition of absolute geometry. The result is used in the proof of the "Euclidean Parallel Theorem" (Th.68).

Theorem 64 *Let $\{Q^x_t\}$ and $\{R^{x'}_{t'}\}$ be two indexed classes of parallels in Σ with $Q^0 = R^0$. If the time scales of Q^0 and R^0 are related by a linear transformation; that is, if for all $t \in \mathbb{R}$, there are real constants c, d such that*

$$Q^0_t = R^0_{t'} = R^0_{c+dt}$$

then the two classes of parallels are of the same type (and conversely).

Proof To prove the direct proposition, we consider any real number a. Then the previous theorem implies that, for all $t \in \mathbb{R}$,

$$(F_{Q^0 Q^a} \circ F_{Q^a Q^0})^*(Q^0_t) = Q^0_{t+2a} \quad \text{and} \quad (F_{R^0 R^{ad}} \circ F_{R^{ad} R^0})^*(R^0_{c+dt}) = R^0_{c+d(t+2a)}$$

and hence Theorem 49 implies that the paths Q^a and R^{ad} coincide at all of their events, so $Q^a = R^{ad}$. But a was arbitrary, so each parallel of one class is also a parallel of the other class and hence the two classes of parallels are of the same type, which establishes the direct proposition.

If the two indexed classes of parallels are of the same type with $Q^0 = R^0$, then there is a real number c such that the event $Q^0_0 \in Q$ is also indexed as $R^0_c \in R$ and there is a positive real number d such that $Q^1 = R^d$. Now for the special case of a dyadic number t, Theorem 60 implies that $Q^{t/2} = R^{dt/2}$ and Theorem 63 implies the stated relation. The extension to the reals is a consequence of the Continuity Theorem (Th.12). $\qquad q.e.d.$

The next theorem shows that for any given path, all time scales of the same type are determined to within an arbitrary strictly increasing linear transformation. In the axiomatic system of Szekeres, this property was regarded as an axiom and was called the "Axiom of Standard Time" (Szekeres, 1968, Axiom A.8). In the present system, the property is a consequence of the Axiom of Isotropy (Axiom S).

Theorem 65 (Affine Invariance of Divergent and Convergent Time Scales)
Let Σ_1 and Σ_2 be collinear sets which meet in (at least) one path and whose temporal order relations conform on that path. Let $\{S^\alpha : S^\alpha \in \Sigma_1, \ \alpha \in \mathbb{R}\}$ and $\{U^\alpha : U^\alpha \in \Sigma_2, \ \alpha \in \mathbb{R}\}$ be indexed classes of parallels of the same type.

(i) *If $S^a = U^b$, then there are real constants c, d, k such that for all real x,*

$$S^a_{c+kx} = U^b_{d+x} \ .$$

(ii) *Also, if $\Sigma_1 = \Sigma_2$, then for all real x and y, the coordinate systems are related by the equation*

$$S^{a \pm ky}_{c+kx} = U^{b+y}_{d+x} \ ,$$

the upper and lower sign being chosen according as to whether the left side of S^a in Σ_1 is the left, or right, side of U^b in Σ_2.

Proof In this proof there are actually two temporal order relations — one for Σ_1 and one for Σ_2. If the two collinear sets are distinct (as could be the case for (i)) then the two order relations conform on the common path $S^a = U^b$ (and the only order relations which will be used are between events of the same collinear set). If $\Sigma_1 = \Sigma_2$ then the Collinear Set Theorem (Th.36) implies that there is no distinction between the two order relations.

(i) The proof given applies to the case of convergent parallels — a similar proof applies to the case of divergent parallels. Take events S^a_0, S^a_x, S^a_y with $S^a_0 < S^a_x < S^a_y$. By the Existence Theorem (Th.54) there are paths $R \in \Sigma_1 \cap SPR[S^a_0]$, $T \in \Sigma_2 \cap SPR[S^a_0]$ such that

$$(F_{S^a R} \circ F_{R S^a})^*(S^a_x) = (F_{S^a T} \circ F_{T S^a})^*(S^a_x) = S^a_y \ ,$$

where the modified record functions are defined with respect to the sense of direction which applies for each collinear set respectively.

The Axiom of Isotropy (Axiom S) implies the existence of a mapping θ with invariant path $S^a \ (= U^b)$ such that $\theta : \ R \to T$ and, since the Collinear Set Theorem 36 implies that Σ_1 is uniquely determined by S^a and R, while Σ_2 is uniquely determined by $U^b \ (= S^a)$ and T, it follows that $\theta : \Sigma_1 \to \Sigma_2$.

By an argument similar to that for the Reflection Mapping Theorem (Th.45), the isotropy mapping θ induces a bijection (independent of the event S_a^0) from Σ_1 to Σ_2. This bijection maps signal functions to signal functions and pairs of parallel paths to pairs of parallel paths of the same type.

For any event $S_c^a \in S^a$ there is some event $U_d^b \in U^b$ such that

$$S_c^a = U_d^b \tag{1}$$

and for an event $U_{d+2}^b \in U^b$ there is a positive real number k such that

$$S_{c+2k}^a = U_{d+2}^b \ . \tag{2}$$

Thus $\theta : U^{b+1} \to S^{a+k}$ and so, by an argument similar to that for the preceding theorem, $\theta : U^{b+y} \to S^{a+ky}$ (for any dyadic number k). The result (i) follows by Theorem 63 and the Continuity Theorem (Th.12).

(ii) If $\Sigma_1 = \Sigma_2$, then according as to whether the left side of S^a is the left, or right, side of U^b, Theorem 49 implies that, for all real y,

$$S^{a \pm ky} = U^{b+y}$$

and so by (i) and the relations given by Theorem 63 it follows that, for all real x and y,

$$S_{c+kx}^{a \pm ky} = U_{d+x}^{b+y} \ .$$

$$q.e.d.$$

7.4 Automorphisms of a collinear set of paths

We will now apply reflection operations to a collinear set of paths. By composing these reflection operations we can obtain mappings called "space translations" and "pseudo–rotations". Finally by composing four of these mappings, we can generate "time translation" mappings.

We first consider compositions of reflections which lead to "spacelike translations" if we compose reflections in two mutually parallel paths, or to "pseudo–rotations" if we compose reflections in two distinct paths which meet at some event. These mappings are discussed in more detail in the following Section 7.5.

Theorem 66 (Mapping of an Indexed Class of Parallels)
Let Σ be a collinear set and let $\{Q^\alpha : \alpha \in \mathbb{R}\}$ and $\{W^\beta : \beta \in \mathbb{R}\}$ be indexed classes of divergent and convergent parallels in Σ such that, for some Q^a and W^d,

$$Q^a = W^d \ .$$

Let V be a path in Σ such that either:

  (i) V meets $Q^a(= W^d)$ at some event, or

  (ii) $V \vee Q^a(= W^d)$, or

  (iii) $V \wedge Q^a(= W^d)$.

Then there is a mapping

$$\phi : \ \Sigma \longrightarrow \Sigma \quad \text{such that} \quad \phi(Q^a) = \phi(W^d) = V$$

and the indexed classes of parallels are mapped onto indexed classes of parallels of the same type. That is, the indexed class of divergent parallels $\{Q^\alpha : \ \alpha \in \mathbb{R}\}$ is mapped onto an indexed class of divergent parallels $\{U^\alpha : \ \alpha \in \mathbb{R}\}$ and this defines a mapping of coordinate systems

$$\phi : \ Q_y^x \to U_y^x \ .$$

A similar result applies to the indexed class of convergent parallels $\{W^\beta : \beta \in \mathbb{R}\}$.

Remark   A mapping corresponding to Case (i) is called a *pseudo–rotation*, while mappings corresponding to Cases (ii) and (iii) are called *spacelike translations*. These mappings preserve the sense of direction assigned to Σ.

Proof   By Theorems 51 and 59, there is a path T mid–way between $Q^a(= W^d)$ and V. We use the symbol ψ (rather than θ) to denote a reflection mapping which leaves the path T invariant. Then the Reflection Mapping Theorem (Th.45) implies that

$$F_{Q^bQ^c}^+ = \psi \circ F_{\psi(Q^b)\psi(Q^c)}^- \circ \psi \quad \text{and} \quad F_{Q^bQ^c}^- = \psi \circ F_{\psi(Q^b)\psi(Q^c)}^+ \circ \psi \qquad (1)$$

where right and left signal functions have been interchanged because ψ is a reflection mapping. Since we are looking for a mapping which will send right and left signal functions onto right and left signal functions, respectively, we shall compose two reflection mappings. Thus we next define a reflection mapping θ, which is a reflection in $\psi(Q^a)(= V)$ and, as above, for any paths R, S in Σ

$$F_{RS}^+ = \theta \circ F_{\theta(R)\theta(S)}^- \circ \theta \quad \text{and} \quad F_{RS}^- = \theta \circ F_{\theta(R)\theta(S)}^+ \circ \theta \ . \qquad (2)$$

If we now define

$$\phi := \theta \circ \psi$$

and combine equations (1) and (2), we obtain

$$F_{Q^bQ^c}^+ = \phi^{-1} \circ F_{\phi(Q^b)\phi(Q^c)}^+ \circ \phi \quad \text{and} \quad F_{Q^bQ^c}^- = \phi^{-1} \circ F_{\phi(Q^b)\phi(Q^c)}^- \circ \phi \qquad (3)$$

130

since $\phi^{-1} = (\theta \circ \psi)^{-1} = \psi^{-1} \circ \theta^{-1} = \psi \circ \theta$. We have already noted that $\{\psi(Q^\alpha)\}$ is a class of parallels of the same type as $\{Q^\alpha\}$, and similarly $\{\phi(Q^\alpha)\}$ is a class of parallels of the same type as $\{Q^\alpha\}$. The Reflection Mapping Theorem (Th.45(ii)) applied to ψ and θ (and their composition ϕ) implies that we can define a set of parallels $\{U^\alpha\}$ with events indexed such that

$$U_y^x := \phi(Q_y^x) . \tag{4}$$

Accordingly, equations (3) imply that

$$U_{y+b-c}^b = F_{U^bU^c}^+(U_y^c) \quad \text{and} \quad U_{y-b+c}^b = F_{U^bU^c}^-(U_y^c), \tag{5}$$

and hence

$$U_{y-2b+2c}^b = (F_{U^bU^c} \circ F_{U^cU^b})^*(U_y^b) . \tag{6}$$

Equations (4), (5) and (6) show that with the indexing specified by (4), the set $\{U^\alpha\}$ is an indexed class of parallels of the same type and by (4),

$$\phi(Q^a) = U^a = V .$$

q.e.d.

In the next theorem we compose two spacelike reflections and two pseudo–rotations to generate a "time translation" mapping.

Theorem 67 (Time Translation)
Let Σ be a collinear set and let $\{Q_y^\alpha\}$ and $\{R_z^\beta\}$ be coordinate systems of divergent and convergent type (respectively) with

$$Q^0 = R^0 \quad and \quad Q_0^0 = R_0^0 .$$

For any $Q_a^0 \in Q^0$ with $Q_0^0 < Q_a^0 < Q_1^0$, there is a bijection

$$\begin{array}{rcl} \tau : & \Sigma & \to \Sigma \\ & Q_x^\alpha & \mapsto Q_{a+kx}^{k\alpha} \\ & R_y^\beta & \mapsto R_{b+ly}^{l\beta} \end{array} ,$$

where k, l, b are positive real numbers. In particular

$$\tau : Q^\alpha \mapsto Q^{k\alpha}, \ R^\beta \mapsto R^{l\beta}, \ Q_x^0 \mapsto Q_{a+kx}^0, \ R_y^0 \mapsto R_{b+ly}^0 .$$

Furthermore, for any non-negative integer n,

$$Q_{a(1+k+\cdots+k^n)}^0 = R_{b(1+l+\cdots+l^n)}^0 .$$

Remark   The mapping τ is called a *time translation mapping* and has the inverse mapping

$$\tau^{-1} : \Sigma \rightarrow \Sigma, \ Q_x^\alpha \mapsto Q_{-a/k+x/k}^{\alpha/k}, \ R_y^\beta \mapsto R_{-b/l+y/l}^{\beta/l} \ .$$

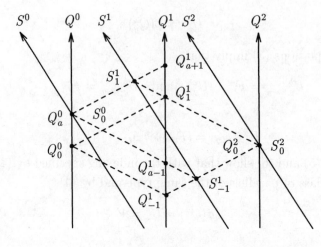

Figure 81

Proof   (see Figure 81). Since $Q_0^0 < Q_a^0 < Q_1^0$, it follows that

$$Q_{-1}^1 < Q_{a+1}^1 \quad \text{and} \quad Q_{a-1}^1 < Q_1^1 \ ,$$

so by Theorem 54 there is a path T in Σ such that

$$(F_{Q^1 T} \circ F_{TQ^1})^*(Q_{-1}^1) = Q_{a-1}^1 \tag{1}$$

and

$$(F_{Q^1 T} \circ F_{TQ^1})^*(Q_{a+1}^1) = Q_1^1 \ . \tag{2}$$

We will now define a coordinate system $\{S_\gamma^\alpha\}$ (it is immaterial whether we choose divergent or convergent parallels) such that $S^1 = T$, and

$$S_{-1}^1 := F_{S^1 Q^1}^+(Q_{-1}^1) \quad \text{and} \quad S_1^1 := F_{S^1 Q^1}^+(Q_{a+1}^1) \ ;$$

whence from (1) and (2),

$$Q_{a-1}^1 = F_{Q^1 S^1}^-(S_{-1}^1) \quad \text{and} \quad Q_1^1 = F_{Q^1 S^1}^-(S_1^1) \ ,$$

and so by Theorem 63 and Theorem 48

$$F^+_{Q^2 S^1}(S^1_{-1}) = Q^2_0 \quad \text{and} \quad F^+_{Q^0 S^1}(S^1_1) = Q^0_a \quad \text{and}$$
$$F^-_{Q^0 S^1}(S^1_{-1}) = Q^0_a \quad \text{and} \quad F^-_{Q^2 S^1}(S^1_1) = Q^2_0 \quad .$$

These four relations imply that the paths $S^0, S^2 \in \{S^\alpha\}$ can be chosen in accordance with Theorem 63 so that S^0 meets Q^0 at $Q^0_a(= S^0_0)$ and S^2 meets Q^2 at $Q^2_0(= S^2_0)$.

By the previous theorem, there are spacelike translation mappings $\phi_1, \phi_3 : \Sigma \to \Sigma$ and pseudo–rotation mappings $\phi_2, \phi_4 : \Sigma \to \Sigma$ which send classes of parallels onto classes of parallels of the same type and preserve the sense of direction assigned to Σ such that

$$\phi_1 : Q^0 \to Q^2 \quad \text{and} \quad Q^0_0 \to Q^2_0$$
$$\phi_2 : Q^2 \to S^2 \quad \text{and} \quad Q^2_0 \to S^2_0 = Q^2_0$$
$$\phi_3 : S^2 \to S^0 \quad \text{and} \quad S^2_0 \to S^0_0$$
$$\phi_4 : S^0 \to Q^0 \quad \text{and} \quad S^0_0 \to Q^0_a = S^0_0 \quad .$$

We now define the *time translation mapping*

$$\tau := \phi_4 \circ \phi_3 \circ \phi_2 \circ \phi_1 \tag{3}$$

and so by the above relations,

$$\tau : Q^0 \to Q^0 \quad \text{and} \quad Q^0_0 \to Q^0_a \quad . \tag{4}$$

Also, by the previous theorem,

$$\tau : \{Q^\alpha\} \to \{Q^\alpha\}, \ \{R^\beta\} \to \{R^\beta\} \ .$$

However we can not assert that τ sends each parallel onto itself; we can only assert that τ sends the right side of Q^0 onto itself (and the left side of Q^0 onto itself). Accordingly, there are real positive numbers k and l such that

$$\tau : Q^1 \to Q^k, \ R^1 \to R^l \ , \tag{5}$$

so by Theorem 65

$$\tau : Q^\alpha \to Q^{k\alpha}, \ R^\beta \to R^{l\beta} \ . \tag{6}$$

There is some positive real number b such that $Q^0_a = R^0_b$. Now for arbitrary x and y,

$$Q^0_x = \left(F^-_{Q^0 Q^{x/2}} \circ F^+_{Q^{x/2} Q^0} \right) (Q^0_0) \quad \text{and} \quad R^0_y = \left(F^-_{R^0 R^{y/2}} \circ F^+_{R^{y/2} R^0} \right) (R^0_0) \ ,$$

and the Reflection Mapping Theorem (Th.45) implies that signals are mapped onto signals, so it follows from (4) and (6) that

$$\tau : Q_x^0 \mapsto Q_{a+kx}^0 , \quad R_y^0 \mapsto R_{b+ly}^0 .$$

By Theorem 63 and the Reflection Mapping Theorem,

$$\tau : \ Q_{x+\alpha}^\alpha = F_{Q^\alpha Q^0}^+ (Q_x^0) \ \mapsto \ F_{Q^{k\alpha} Q^0}^+ (Q_{a+kx}^0) = Q_{a+kx+k\alpha}^{k\alpha}$$

and similar considerations apply to convergent parallels, so

$$\tau : \ Q_x^\alpha \mapsto Q_{a+kx}^{k\alpha} , \quad R_y^\beta \mapsto R_{b+ly}^{l\beta} .$$

Since $Q_0^0 = R_0^0$, it follows that for any positive integer n,

$$\tau^n : \ Q_0^0 = R_0^0 \mapsto Q_{a(1+k+\cdots+k^n)}^0 = R_{b(1+l+\cdots+l^n)} .$$

$$q.e.d.$$

7.5 Uniqueness of parallelism and linearity of modified signal functions

The culmination of the present theory of parallelism is in the next theorem where we show that there is only one type of parallel; that is, there is no distinction between convergent parallels and divergent parallels. It turns out (in Chapters 9 and 10) that the only space–time satisfying our axioms is the usual Minkowski space–time, which shares the property of uniqueness of parallelism with Euclidean geometry. Hyperbolic geometry, which is the other absolute geometry, does not have this property of uniqueness of parallelism; likewise the de Sitter space–times do not have this property.

Some authors describe one of these de Sitter space–times as cohyperbolic geometry (Yaglom et al (1964), Yaglom (1969)) or as locally-timelike hyperbolic geometry (Busemann (1967)) since it can be represented as the exterior of a closed ellipsoid in four-dimensional projective space with paths being straight lines which intersect the interior. The other de Sitter space–times are coverings of this space–time.

The property of uniqueness of parallelism implies that there is one type of time scale, called a *"natural time scale"* and one type of coordinate system called a *"natural coordinate system"*. The spacelike and timelike translation mappings can then be re-expressed in a more convenient form in Theorems 69 and 70. We conclude the chapter with Theorem 71 in which we establish the linearity of modified signal functions and obtain the affine description of paths and optical lines with respect to natural coordinate systems. At this stage considerations of rectilinear kinematics can begin and the discussion forms the content of the following Chapter 8. For readers who wish to proceed directly to the full $3 + 1$–dimensional Minkowski space–time, it is possible to proceed directly to Chapter 9 without any loss of logical continuity.

Theorem 68 ("Euclidean" Parallel Theorem)
Let Σ be a collinear set containing distinct paths Q, S and let W_x be an event of Σ. Then

$$S \vee (Q, W_x) \iff S \wedge (Q, W_x) .$$

That is, given any path and any event, there is exactly one parallel to the given path through the given event and there is no distinction between divergent parallels and convergent parallels.

Proof Let $\{Q_y^\alpha\}$ and $\{R_z^\beta\}$ be divergent and convergent coordinate systems as in the previous theorem with $Q = Q^0 = R^0$ and $Q_0^0 = R_0^0$.

(i) If there are two coincident parallels

$$Q^a = R^b \qquad (a, b \neq 0),$$

one of each type, then for any dyadic number $d = n/2^m$,

$$Q^{ad} = R^{bd}$$

and hence the two classes of parallels are identical. Thus the two classes of parallels are either identical (apart from indexing), or distinct with no common membership apart from $Q = Q^0 = R^0$. In (ii) (below) we will establish the existence of Q^a and R^b such that $Q^a = R^b$.

(ii) By the previous theorem, there can be a single fixed event $Q_{a/(1-k)}^0 = R_{b/(1-\ell)}^0$ if and only if k and ℓ are both greater than 1 or both less than 1. Thus there are three cases: (a) $k > 1$ and $\ell > 1$, (b) $k < 1$ and $\ell < 1$ and (c) $k = 1$ and $\ell = 1$ which will be considered separately.

Case (a) $k > 1$ and $\ell > 1$. We define a new divergent coordinate system $\{S_\gamma^\alpha\}$ and a new convergent coordinate system $\{T_\delta^\beta\}$ in Σ such that

$$S^0 = Q^0 \ , \ T^0 = R^0 \ , \ S_0^0 = Q_{a/(1-k)}^0 \ , \ T_0^0 = R_{b/(1-\ell)}^0 \ , \ \text{and}$$

$$(F_{S^0 S^{1/2}} \circ F_{S^{1/2} S^0})^*(S_0^0) = S_1^0 = T_1^0 = (F_{T^0 T^{1/2}} \circ F_{T^{1/2} T^0})^*(T_0^0)$$

(where the event $S_1^0(=T_1^0)$ is any chosen event after the fixed event $S_0^0(=T_0^0)$). The time translation mapping τ has properties which can be expressed more directly with respect to the indexed classes $\{S^\alpha\}$ and $\{T^\beta\}$: thus

$$\tau: \ S^a \mapsto S^{ka} \ , \ T^\beta \mapsto T^{\ell\beta} \ , \ S_y^0 \mapsto S_{ky}^0 \ , \ T_z^0 \mapsto T_{\ell z}^0 \ , \tag{1a}$$

the fixed event is $S_0^0(=T_0^0)$ and also, since $S_1^0 = T_1^0$,

$$S_{k^n}^0 = T_{\ell^n}^0 \ . \tag{1b}$$

By Theorem 62 the set of events of Q is order–isomorphic to the sets of real numbers which index the events of S^0 and T^0, so there is a strictly monotonic increasing bijection g from the real indices of S^0 to the real indices of T^0; that is,

$$g: \ S^0 \rightarrow T^0$$

$$y \mapsto z \qquad \text{if and only if} \qquad S_y^0 = T_z^0 \ . \tag{2}$$

We will now define the function

$$h(y) := g(y+1) - g(y) \tag{3}$$

which is continuous since g is continuous. If $S^{1/2}$ crosses $T^{h(y)/2}$ at the event $F_{S^{1/2}S^0}(S_y^0)$ then, by Theorem 49 and the Indexing Theorem (Th.63)

$$f_{S^0S^{1/2}} \circ f_{S^{1/2}S^0}(S_y^0) = S_{y+1}^0$$

and

$$f_{T^0T^{h(y)/2}} \circ f_{T^{h(y)/2}T^0}(T_{g(y)}^0) = T_{g(y+1)}^0 \ .$$

If the classes of parallels are distinct then h is a strictly monotonic function, since otherwise $S^{1/2}$ would cross some convergent parallel $T^{h(y)/2}$ at two distinct events, which would contradict the Axiom of Uniqueness (Axiom I3): furthermore h is increasing for $y = 1/2$, so h is a strictly monotonic increasing function. Next we will show that h is an unbounded function by showing that, for each real number $\beta > 1/2$, the path $S^{1/2}$ crosses T^β: suppose the contrary; that is, suppose there is some T^β (with $\beta > 1/2$) which $S^{1/2}$ does not cross. Now $T^\beta \wedge S^0$ and T^β does not meet $S^{1/2}$ so $T^\beta \wedge S^{1/2}$. By symmetry, $S^{1/2} \wedge T^\beta$ and since $T^\beta \wedge S^0$, the transitivity of parallelism implies $S^{1/2} \wedge S^0$. By (i) this would imply that the two classes of parallels are identical. Thus h is unbounded, so there is some integer n_a such that:

$$\text{for all real } y > n_a \ , \quad h(y) > 2 \ . \tag{4}$$

Since $k > 1$, the two sequences $(k^m : m = 1, 2, \cdots)$ and $(k^{m+1} - k^m : m = 1, 2, \cdots)$ are both unbounded so there is some integer n_b such that:

$$\text{for all integers } m > n_b, \quad k^m > n_a \quad \text{and} \quad k^{m+1} - k^m > 2. \tag{5}$$

Let $K(m)$ be the largest non-negative integer such that

$$K(m) < k^{m+1} - k^m. \tag{6}$$

Let $n := max\{n_a, n_b\}$. Then for any integer $m > n$,

$$\frac{k^{m+1} - k^m}{2} < K(m) < k^{m+1} - k^m \tag{7}$$

By (1b), (2) and (7), for $m > n$,

$$\ell^{m+1} = g(k^{m+1})$$
$$> g(k^m + K(m))$$

since g is a strictly monotonic increasing function and, from (3),

$$\ell^{m+1} > h(k^m + K(m) - 1) + h(k^m + K(m) - 2) + \cdots + h(k^m) + g(k^m).$$

Also (4) and (5) imply that $h(k^m) > 2$ and h is a strictly monotonic increasing function, so

$$\ell^{m+1} > 2K(m) + g(k^m)$$
$$> k^{m+1} - k^m + \ell^m \quad \text{by (1b), (2) and (7)}.$$

Therefore, for any integer $m > n$,

$$\ell^{m+1} - \ell^m > k^{m+1} - k^m$$

and since $k > 1$ and $\ell > 1$,

$$\left(\frac{\ell}{k}\right)^m > \frac{k-1}{\ell-1} \quad \text{for all } m > n$$

which implies that

$$k < \ell. \tag{8a}$$

The direction of time has been specified by convention. If the opposite direction is chosen, then the previous classes of convergent and divergent parallels become classes of divergent and convergent parallels respectively with matching constants $k' = \ell$ and $\ell' = k$ and the same argument shows that

$$k' < \ell' \ ,$$

that is

$$\ell < k \ . \tag{8b}$$

The inequalities (8a,b) are contradictory so we conclude, by (i) above, that the two classes of parallels are not distinct.

Case (b) $k < 1$ and $\ell < 1$. Consider the mapping $\tau^* := \tau^{-1}$ and reverse the direction of time. Then the previous classes of convergent and divergent parallels become classes of divergent and convergent parallels respectively with matching constants $k^* = \ell^{-1} > 1$ and $\ell^* = k^{-1} > 1$, which is simply an instance of the previous Case (a).

Case (c) $k = 1$ and $\ell = 1$. By the previous theorem,

$$\tau : \ Q_0^0 = R_0^0 \ \mapsto \ Q_a^0 = R_b^0 \qquad \text{and} \qquad \tau : \ Q_a^0 = R_b^0 \ \mapsto \ Q_{2a}^0 = R_{2b}^0$$

so by Theorem 49, the paths $Q^{a/2}$ and $R^{b/2}$ meet at the two events $f^+{}_{Q^{a/2}Q^0}(Q_0^0)$ and $f^+{}_{Q^{a/2}Q^0}(Q_a^0)$. The Axiom of Uniqueness (Axiom I3) implies that

$$Q^{a/2} = R^{b/2}$$

so, by (i) above, the two classes of parallels are identical.

We have seen that in each of the three cases the two classes of parallels are not distinct. Thus we have demonstrated the "uniqueness of parallelism". $\qquad q.e.d.$

An immediate consequence of the previous theorem and Theorem 65 is that there is no distinction between divergent and convergent time scales and the time scales are defined, independently of the collinear set, to within an arbitrary linear transformation. Within a particular collinear set the time scales are determined to within an arbitrary strictly increasing linear transformation and are called *natural time scales*: similarly there is no distinction between divergent and convergent coordinate systems which, from now on, will be called *natural coordinate systems*.

138

Theorem 69 (Space Displacement Mapping)

Let $\{Q_t^x\}$ be a natural coordinate system in a collinear set Σ. Given any real numbers a and b, there is a bijection

$$\delta : \Sigma \to \Sigma$$
$$Q_t^x \mapsto Q_t^{x-a+b}$$

and, for all R in Σ,

$$\delta(R) \parallel R \ .$$

Furthermore, for any natural coordinate system $\{U_{t'}^{x'}\}$ in Σ there are real constants c and d such that

$$\delta : \ U_{t'}^{x'} \to U_{d+t'}^{c+x'} \ .$$

The mapping δ is called a "space displacement mapping".

Proof The case $a = b$ is trivial, so from now on we assume that $a \neq b$. This proof is based on the proof of Theorem 66: accordingly, we let $V := Q^b$ and we let $T := Q^{(a+b)/2}$ and we consider two reflection mappings ψ and θ where ψ is a reflection of Σ in T and θ is a reflection of Σ in V. Thus

$$\psi : \ Q_t^x \mapsto Q_t^{a+b-x} \quad \text{and} \quad \theta : \ Q_t^x \mapsto Q_t^{2b-x} \ ,$$

whence, if we define $\delta := \phi := \theta \circ \psi$,

$$\delta : \ Q_t^x \mapsto Q_t^{x-a+b} \ . \tag{1}$$

Given any path R in Σ, since $a \neq b$, (1) implies that there is no event at which R and $\delta(R)$ meet, so by the previous theorem,

$$\delta(R) \parallel R \quad \text{and} \quad \delta(R) \neq R \tag{2}$$

and there is no fixed event with respect to the mapping δ. Theorem 63 implies that for any real numbers t', x', y' with $x' < y'$,

$$U_{t'}^{x'} \ \sigma \ U_{t'-x'+y'}^{y'} \quad \text{and} \quad U_{t'}^{y'} \ \sigma \ U_{t'-x'+y'}^{x'} \ . \tag{3}$$

These relations correspond to right and left modified signal functions respectively. Since δ is a composition of reflection mappings which preserve signal relations (by Theorem 45),

$$\delta(U_{t'}^{x'}) \ \sigma \ \delta(U_{t'-x'+y'}^{y'}) \quad \text{and} \quad \delta(U_{t'}^{y'}) \ \sigma \ \delta(U_{t'-x'+y'}^{x'}) \tag{4}$$

in accordance with the results of Theorem 63 so we can define a natural coordinate system $\{V_{t'}^{x'}\}$ in Σ such that

$$V_{t'}^{x'} := \delta(U_{t'}^{x'}) \; .$$

By (2), there is some V^β such that

$$U^0 = V^\beta \; ,$$

and so by the affine invariance of time scales (Theorem 65) there are real constants c, d, k such that for all real t' and x',

$$U_{d+kt'}^{c+kx'} = V_{t'}^{x'} \quad (= \delta(U_{t'}^{x'})) \; . \tag{5}$$

If $k \neq 1$ we can choose $t' = (|\, c\, | - d)/(k - 1)$ and then

$$U_{t'}^0 \; \sigma \; \delta(U_{t'}^0) \; ,$$

which, by Theorem 63, is a contradiction by (1). Thus $k = 1$; whence (5) becomes

$$\delta(U_{t'}^{x'}) = U_{d+t'}^{c+x'} \; .$$

q.e.d.

Theorem 70 (Time Translation Mapping)
Let $\{U_t^x\}$ be a natural coordinate system in Σ with events $U_c^b, U_d^b \in U^b$ such that $U_c^b < U_d^b$. There is a bijection

$$\tau : \quad \Sigma \to \Sigma$$
$$U_t^x \mapsto U_{t-c+d}^x \; .$$

The mapping τ is called a "time displacement mapping".

Proof This proof is based on the proof of Theorem 67. We define a natural coordinate system $\{Q_t^x\}$ in Σ such that, for some real number a with $0 < a < 1$,

$$Q^0 := U^b \; , \quad Q_0^0 := U_c^b \; , \quad Q_a^0 := U_d^b \; .$$

By Theorem 65

$$Q_t^x = U_{c+kt}^{b+kx} \quad \text{where} \quad k = (d - c)/a \; . \tag{1}$$

140

The space displacement mappings ϕ_1 and ϕ_3 (of the proof of Theorem 67) are such that

$$\phi_1 : Q_0^0 \mapsto Q_0^2 \quad \text{and} \quad \phi_3 : Q_0^2 \mapsto Q_a^0 \ ,$$

so by the previous theorem,

$$\phi_1 : Q_t^x \mapsto Q_t^{2+x} \quad \text{and} \quad \phi_3 : Q_t^x \mapsto Q_{t+a}^{-2+x} \ .$$

We now define a *time displacement mapping* $\tau^* := \phi_3 \circ \phi_1$ (whose definition is different from the τ of Theorem 67) and then

$$\tau^* : Q_t^x \mapsto Q_{t+a}^x \ ,$$

whence from (1),

$$\tau^* : U_{c+kt}^{b+kx} \mapsto U_{d+kt}^{b+kx} \ ,$$

which is equivalent to

$$\tau^* : U_t^x \mapsto U_{t-c+d}^x$$

which is the required mapping. $\hspace{3cm}$ *q.e.d.*

In the next theorem we show that modified signal functions are linear and that paths (as well as optical lines) are represented by linear equations. Optical lines have "rectilinear velocities" with unit magnitude, while paths have rectilinear velocities whose magnitude is less than unity.

Theorem 71 (Linearity of Modified Signal Functions)

(i) *If Q and R are paths in Σ with natural time scales, then*

$$F_{QR}^+(R_t) = Q_{at+b} \quad \text{and} \quad F_{QR}^-(R_t) = Q_{ct+d} \ ,$$

where a, b, c, d are constants and both a and c are positive. Furthermore

$$F_{RQ}^+(Q_t) = R_{(t-b)/a} \quad \text{and} \quad F_{RQ}^-(Q_t) = Q_{(t-d)/c} \ .$$

(ii) *With respect to a natural coordinate system in Σ, the set of events of any path may be represented by the linear equation*

$$x = x_0 + vt \ , \quad \text{where} \quad |v| < 1 \ .$$

Conversely any set of events which satisfy this equation lie on a path. Optical lines satisfy the same equation but with $v = +1$ or $v = -1$, where the sign depends on whether the line is a right optical line or a left optical line. Thus the set of paths and optical lines in Σ has an affine structure and the concept of parallelism applies to optical lines as well as to paths.

Remark The constant v is called the *rectilinear velocity* of the path with respect to the coordinate system.

Proof (i) If Q and R meet at no event, or if they are permanently coincident, they are parallel and the result is a special case of Theorems 63 and 65, otherwise Q and R meet at some event by Theorem 68.

We now define natural coordinate systems $\{S_t^x\}$ and $\{U_{t'}^{x'}\}$ in Σ such that

$$S^0 = Q \ , \ U^0 = R \quad \text{and} \quad S_0^0 = U_0^0 \ .$$

For any real number a, the paths S^a and U^0 are not parallel, so they meet (and cross) at some event

$$S_b^a = U_c^0 \ . \tag{1}$$

Let δ and τ be space and time translations, respectively, as in the preceding two theorems, such that

$$\delta : \ S_t^x \mapsto S_t^{x+a} \quad \text{and} \quad \tau : \ S_t^x \mapsto S_{t+b}^x \ .$$

Consequently we can define a mapping $\lambda := \delta \circ \tau = \tau \circ \delta$ such that

$$\lambda : \ S_t^x \mapsto S_{t+b}^{x+a} \ . \tag{2}$$

Since τ is a composition of two space displacement mappings, λ is a composition of three space displacement mappings, so by Theorem 69 and (1),

$$\lambda : \ U_{t'}^{x'} \to U_{t'+c}^{x'} \ . \tag{3}$$

Since δ and τ are bijections, λ is a bijection and so, for any integer n,

$$\lambda^n : \ S_0^0 = U_0^0 \mapsto S_{nb}^{na} = U_{nc}^0 \ .$$

Now a was arbitrary, so if we choose any positive integer m and substitute $a/2^m$ for a wherever a appears, we find that for any positive integer m and for any integer n,

$$S_{bn/2^m}^{an/2^m} = U_{cn/2^m}^0 \ ;$$

that is, for any dyadic number p,

$$S_{bp}^{ap} = U_{cp}^0 \ . \tag{4}$$

Consequently by Theorem 63

$$F_{U^0 S^0}^+ (S_{bp-ap}^0) = U_{cp}^0 \quad \text{and} \quad F_{U^0 S^0}^- (S_{bp+ap}^0) = U_{cp}^0$$

and, since signal functions are continuous by Theorem 43, it follows that for all real t,

$$F_{U^0 S^0}^+(S_t^0) = U_{ct/(b-a)}^0 \quad \text{and} \quad F_{U^0 S^0}^-(S_t^0) = U_{ct/(b+a)}^0 \ .$$

That is, the modified signal functions $F_{U^0 S^0}^+$ and $F_{U^0 S^0}^-$ are linear strictly increasing functions and therefore Theorem 65 implies that F_{RQ}^+ and F_{RQ}^- are linear strictly increasing functions which can be written in the general form

$$F_{RQ}^+(Q_t) = R_{At+B} \quad \text{and} \quad F_{RQ}^-(Q_t) = R_{Ct+D} \ ,$$

where A, B, C, D are constants and A and C are positive. By Theorem 48

$$F_{QR}^+(R_t) = Q_{(t-B)/A} \quad \text{and} \quad F_{QR}^-(R_t) = Q_{(t-D)/C} \ .$$

(ii) Now by (i) the modified signal functions between any path V in Σ and the path S^0 are linear, so the corresponding modified record function is linear; that is

$$(F_{S^0 V} \circ F_{V S^0})^* \left(S_w^0\right) = S_{aw+b}^0$$

where a, b are constants. By Theorem 63(iii), the position–time coordinates $(x; t)$ of events $S_t^x \in V$ are given by

$$x = \frac{1}{2}[(aw + b) - w] \quad \text{and} \quad t = \frac{1}{2}[(aw + b) + w]$$

from which we see that x is a linear function of t, namely

$$x = x_0 + vt \ , \quad \text{where} \quad |v| < 1 \tag{5}$$

since the unreachable set is connected (Theorem 13). Conversely any set of events which satisfy (5) lie on a path. $\qquad q.e.d.$

It is now possible to develop the kinematics of rectilinear motion and to discuss the concepts of congruence and synchronism of time scales. This will be the subject matter of the next chapter. For the reader who wishes to proceed directly to the full $3 + 1$–dimensional Minkowski space–time, it is possible to proceed directly to Theorem 80 of the subsequent Chapter 9 without any loss of logical continuity.

8. One–dimensional kinematics

Since a collinear set corresponds to a set of paths in "one-dimensional motion", the relationships between events, paths, and optical lines are most appropriately described in kinematic terms. The results of this chapter provide insights into some properties of Minkowski space–time which distinguish it from the Galilean space–time of Newtonian mechanics. For the reader who wishes to proceed directly to the full $3 + 1$–dimensional Minkowski space–time, it is possible to proceed directly to Theorem 80 of the subsequent Chapter 9 without any loss of logical continuity.

In this chapter we restrict our attention to a collinear set, so all paths have natural time scales and modified signal functions are linear. We will often delete the path symbol where there is no chance of ambiguity; for example, in the next theorem, instead of writing

$$(F_{QS} \circ F_{SQ})^*(Q_x) = Q_{M_{SQ}(x-q)+q} \; ,$$

we shall write

$$(F_{QS} \circ F_{SQ})^*(x) = M_{SQ}(x - q) + q \; .$$

8.1 Rapidity is a natural measure for speed

In this section we define "rapidity" which is a non-dimensional measure of speed. For collinear sets of paths, directed rapidities are composed by simple arithmetic addition, which means that rapidity is a natural measure for speed. The name "rapidity" is due to Robb (1911) who introduced this concept in a different way.

Theorem 72

(i) *If $S \nparallel Q$, there is a positive "constant of the motion" M_{SQ} and a real number q such that*

$$(F_{QS} \circ F_{SQ})^*(x) = M_{SQ}(x - q) + q \; ,$$

where the real number q is such that S coincides with Q at Q_q .

(ii)

$$M_{QS} = (M_{SQ})^{-1} \; .$$

(iii) *If $R \parallel Q$ and $T \parallel S$, then $M_{TR} = M_{SQ}$.*

Remark The constant of the motion M is invariant with respect to affine transformations of natural time scales, by part (iii).

Proof (i) By the previous theorem, both F_{SQ}^+ and F_{QS}^- are linear increasing functions, so their composition $(F_{QS} \circ F_{SQ})^*$ is a linear increasing function. If $Q \parallel\!\!\!/ \ S$, there is some event $Q_q \in Q$ such that S coincides with Q at Q_q and so, by Theorem 49, the record function is of the form

$$(F_{QS} \circ F_{SQ})^*(x) = M_{SQ}(x - q) + q ,$$

where $M_{SQ} > 0$.

(ii) The previous theorem implies that, for any two paths Q,S in Σ, there are constants β_{QS}, q_s^-, α_{SQ}, s_q^+ such that

$$F_{QS}^-(x) = \beta_{QS}x + q_s^- \qquad \text{and} \qquad F_{SQ}^+(x) = \alpha_{SQ}x + s_q^+ ,$$

so by Theorem 48

$$F_{SQ}^-(x) = \beta_{QS}^{-1}x - \beta_{QS}^{-1}q_s^- \qquad \text{and} \qquad F_{QS}^+(x) = \alpha_{SQ}^{-1}x - \alpha_{SQ}^{-1}s_q^+ ;$$

that is

$$\beta_{SQ} = \beta_{QS}^{-1} \qquad \text{and} \qquad \alpha_{QS} = \alpha_{SQ}^{-1} ,$$

whence

$$M_{SQ} = \beta_{QS}\alpha_{SQ} = (\beta_{SQ}\alpha_{QS})^{-1} = M_{QS}^{-1} .$$

(iii) The previous theorem implies that there are constants $\alpha_{QR}, \beta_{RQ}, r_q^-, q_r^+$ such that

$$F_{RQ}^-(x) = \beta_{RQ}x + r_q^- \qquad \text{and} \qquad F_{QR}^+(x) = \alpha_{QR}x + q_r^+ ,$$

and since the paths Q and R are parallel, $\beta_{RQ}\alpha_{QR} = 1$ so

$$F_{RQ}^-(x) = \alpha_{QR}^{-1}x + r_q^- \qquad \text{and} \qquad F_{QR}^+(x) = \alpha_{QR}x + q_r^+ .$$

Consequently by Theorem 48

$$\begin{aligned}
(F_{RS} \circ F_{SR})^*(R_x) &= F_{RS}^- \circ F_{SR}^+(x) \\
&= F_{RQ}^- \circ F_{QS}^- \circ F_{SQ}^+ \circ F_{QR}^+(x) \\
&= F_{RQ}^- \circ (F_{QS} \circ F_{SQ})^* \circ F_{QR}^+(x) \\
&= M_{SQ}x + \alpha_{QR}^{-1}\left[M_{SQ}(q_r^+ - q) + q\right] + r_q^- .
\end{aligned}$$

Thus we have shown that
$$M_{SR} = M_{SQ} \ ,$$
and similarly, since $T \parallel S$,
$$M_{RT} = M_{RS} \ ,$$
and so by (ii),
$$M_{TR} = (M_{RT})^{-1} = (M_{RS})^{-1} = M_{SR} = M_{SQ} \ .$$

$$q.e.d.$$

Given any paths Q, S in Σ such that $S \parallel Q$, Theorem 63 shows that

$$(F_{QS} \circ F_{SQ})^*(x) = x + 2d \ ,$$

where d is a real constant. The constant of the motion M_{SQ} is 1, and is therefore not shown explicitly. The results of the preceding theorem (with the exception of (i)) apply trivially to the case where $S \parallel Q$.

Given any two paths Q, S in Σ we define the *directed rapidity* of S relative to Q to be

$$r_{SQ} := \frac{1}{2} \log_e M_{SQ} \ .$$

Since $M > 0$,

$$-\infty < r < \infty \ .$$

In the case of parallel paths, $M = 1$ and hence $r = 0$. We call $|r_{SQ}|$ the *relative rapidity* of S with respect to Q.

The next theorem is in no way surprising, since rapidity has been defined so that we would have a measure for speed which is unbounded and which is composed by simple arithmetic addition. Also, the previous theorem implies that rapidities are unaltered by arbitrary affine transformations of natural time scales, so the following result is invariant with respect to transformations of natural time scales.

Theorem 73 (Addition Law for Directed Rapidity)
For any paths Q, S, T in Σ,

$$r_{TQ} = r_{TS} + r_{SQ} \ .$$

That is, "rapidity is a natural measure for speed".

146

Proof By Theorem 71 there are real constants β_{QS}, q_s^-; α_{SQ}, s_q^+; β_{ST}, s_t^-; α_{TS}, t_s^- such that

$$F_{QS}^-(x) = \beta_{QS}x + q_s^- \qquad \text{and} \qquad F_{SQ}^+(x) = \alpha_{SQ}x + s_q^+ \qquad \text{and}$$
$$F_{ST}^-(x) = \beta_{ST}x + s_t^- \qquad \text{and} \qquad F_{TS}^+(x) = \alpha_{TS}x + t_s^+$$

By Theorem 48,

$$F_{QT}^-(x) = \beta_{QS}\beta_{ST}x + \beta_{QS}s_t^- + q_s^- \quad \text{and}$$
$$F_{TQ}^+(x) = \alpha_{TS}\alpha_{SQ}x + \alpha_{TS}s_q^+ + t_s^+ \quad,$$

whence, as in part (ii) of the previous theorem,

$$M_{TQ} = \beta_{QS}\beta_{ST}\alpha_{TS}\alpha_{SQ}$$
$$= \beta_{ST}\alpha_{TS}\beta_{QS}\alpha_{SQ}$$
$$= M_{TS}M_{SQ},$$

and taking logarithms of both sides,

$$r_{TQ} = r_{TS} + r_{SQ} .$$

$$q.e.d.$$

If $\{Q_t^x\}$ and $\{S_{t'}^{x'}\}$ are two natural coordinate systems, it follows from Theorem 72(iii) that there is a unique directed rapidity of the (parallel paths of the) coordinate system $\{Q_t^x\}$ with respect to the (parallel paths of the) coordinate system $\{S_{t'}^{x'}\}$.

8.2 Congruence of a collinear set of paths

Given any two paths Q, S in Σ there are constants α_{SQ}, β_{QS}, s_q^+, q_s^- such that

$$F_{SQ}^+(x) = \alpha_{SQ}x + s_q^+ \qquad \text{and} \qquad F_{QS}^-(x) = \beta_{QS}x + q_s^- .$$

We say that Q and S are *congruent* if $\alpha_{SQ} = \beta_{QS}$ which is equivalent to the condition $\alpha_{QS} = \beta_{SQ}$ since $\alpha_{QS} = \alpha_{SQ}^{-1}$ and $\beta_{SQ} = \beta_{QS}^{-1}$.

The word "congruent" has also been used by Milne (1948) in a different sense: here, we use the word "synchronous" (see the following Section 8.3) where Milne used the word "congruent". We have departed in terminology from Milne because the word congruent is descriptive of the idea of equality of time durations and therefore is in closer correspondence with the geometric concept of the same name.

Theorem 74 *Congruence is an equivalence relation on a collinear set of paths.*

Proof By definition, the relation of congruence is a reflexive and symmetric relation.

In order to show that congruence is a transitive relation, we consider three paths Q,R,S in Σ such that Q is congruent to R and R is congruent to S; that is, $\alpha_{RQ} = \beta_{QR}$ and $\alpha_{SR} = \beta_{RS}$. Then by Theorem 48,

$$\alpha_{SQ} = \alpha_{SR}\alpha_{RQ} = \beta_{RS}\beta_{QR} = \beta_{QR}\beta_{RS} = \beta_{QS} ,$$

which shows that Q is congruent to S. *q.e.d.*

If Q and S are paths in Σ which are not congruent we can define a path $T = S$ whose natural time scale is defined such that

$$F_{TS}^+(x) := (\beta_{QS}/\alpha_{SQ})^{\frac{1}{2}}x = F_{TS}^-(x) .$$

Then

$$\alpha_{TQ} = (\beta_{QS}\alpha_{SQ})^{\frac{1}{2}} = M_{SQ}^{\frac{1}{2}} = \beta_{QT} ,$$

from which we see that Q and T are congruent. Since the natural time scale of each path is only determined to within an arbitrary affine transformation, we could further specify the time scales of paths in a particular collinear set, by choosing a given path, say Q in Σ, and specifying that each other path in Σ is congruent to Q. By the preceding theorem, all paths in the collinear set are now congruent to each other. (Since this theorem only applies to one collinear set of paths, we must be careful not to apply the theorem to two or more distinct collinear sets of paths, for this would assume the transitivity of congruence for non-collinear paths).

Given two congruent paths Q,S in Σ such that $Q \parallel S$, their record functions are of the form:

$$(F_{QS} \circ F_{SQ})^*(x) = x + 2d_{SQ} \quad \text{and} \quad (F_{SQ} \circ F_{QS})^*(x) = x + 2d_{QS} ,$$

where the constants d_{SQ} and d_{QS} are called the *directed distance of S relative to Q*, and the *directed distance of Q relative to S*, respectively. The directed distances are defined in terms of the time scales of the paths, and so they are not invariant with respect to transformations of natural time scales.

148

Theorem 75 (Additivity of Directed Distances)

Let Q, S, T be congruent paths in a collinear set. If $Q \parallel S \parallel T$, then

(i) $d_{TQ} = d_{TS} + d_{SQ}$ and

(ii) $d_{QS} = -d_{SQ}$.

Remark It is important to note that, in contrast to the analogous property (Theorem 73) for collinear rapidities, this property only applies if Q,S,T are congruent, since the directed distances are defined in terms of the time scales of the paths.

Proof Part (ii) is a special case of part (i) with $T = Q$, so we only need to prove part (i). Since Q,S,T are congruent, there are constants γ_{SQ}, γ_{TS}, δ_{QS}, δ_{ST} such that

$$F_{SQ}^+(x) = x + \gamma_{SQ} \qquad , \qquad F_{TS}^+(x) = x + \gamma_{TS} \qquad \text{and}$$
$$F_{QS}^-(x) = x + \delta_{QS} \qquad , \qquad F_{ST}^-(x) = x + \delta_{ST} .$$

By Theorem 48,

$$F_{TQ}^+(x) = x + \gamma_{TS} + \gamma_{SQ} \qquad \text{and} \qquad F_{QT}^-(x) = x + \delta_{QS} + \delta_{ST} .$$

Thus

$$(F_{QT} \circ F_{TQ})^*(x) = x + (\delta_{ST} + \gamma_{TS}) + (\delta_{QS} + \gamma_{SQ})$$
$$= x + 2d_{TS} + 2d_{SQ} ,$$

since $\delta_{QS} + \gamma_{SQ} = 2d_{SQ}$ and $\delta_{ST} + \gamma_{TS} = 2d_{TS}$.   *q.e.d.*

8.3 Partitioning a collinear set into synchronous equivalence classes

Given any paths Q, S in Σ we say that Q and S are *synchronous* if

$$F_{SQ}^+(x) = F_{QS}^-(x) \qquad \text{and} \qquad F_{SQ}^-(x) = F_{QS}^+(x) .$$

(One condition implies the other, by Theorem 48).

Theorem 76 (Synchronous Parallel Paths)

The synchronous relation is an equivalence relation on any collinear class of parallels.

Proof By definition, the synchronous relation is reflexive and symmetric. In order to show that the synchronous relation is transitive, we consider three paths Q, S, T in Σ such that $Q \parallel S \parallel T$ and such that the pairs Q, S and S, T are synchronous. It then follows from the definition of directed distance, that

$$F_{SQ}^+(x) = F_{QS}^-(x) = x + d_{SQ} \qquad \text{and} \qquad F_{TS}^+(x) = F_{ST}^-(x) = x + d_{TS} \ .$$

Then, by Theorem 48,

$$\begin{aligned}
F_{TQ}^+(x) = F_{TS}^+ \circ F_{SQ}^+(x) &= x + d_{TS} + d_{SQ} \\
&= x + d_{SQ} + d_{TS} \\
&= F_{QS}^- \circ F_{ST}^-(x) \\
&= F_{QT}^-(x) \ .
\end{aligned}$$

$$q.e.d.$$

Theorem 77 (Synchronous Collinear Sub–SPRAYs)
The synchronous relation is an equivalence relation on any collinear sub–SPRAY.

Proof By definition, the synchronous relation is reflexive and symmetric. To show that it is transitive on the set of paths belonging to a collinear sub–SPRAY, we consider three paths Q,S,T belonging to the sub–SPRAY such that the pairs Q,S and S,T are synchronous. At the event of coincidence, the real index of Q must be the same as the real index of S, which must also be the same as the real index of T, since both pairs are synchronous. Therefore the modified signal functions are of the form:

$$\begin{aligned}
F_{SQ}^+(x) = F_{QS}^-(x) &= \alpha(x - a) + a \qquad \text{and} \\
F_{TS}^+(x) = F_{ST}^-(x) &= \beta(x - a) + a \qquad .
\end{aligned}$$

By Theorem 48,

$$F_{TQ}^+(x) = F_{TS}^+ \circ F_{SQ}^+(x) = \alpha\beta(x - a) + a = F_{QS}^- \circ F_{ST}^-(x) = F_{QT}^-(x) \ .$$

$$q.e.d.$$

We have shown that the synchronous relation is an equivalence relation on any class of parallels and also on any collinear sub–SPRAY: these are the only subsets of paths on which the synchronous relation is an equivalence relation. Thus it is not possible for all the paths of a collinear set to be synchronous. However, all paths of a single chosen collinear sub–SPRAY could be synchronous, and each class of parallels could be synchronous with that member which is contained in the given synchronous collinear sub–SPRAY.

8.4 Kinematic relations and coordinate transformations

Our discussion of rectilinear kinematics reaches its culmination in the next theorem. In part (i) we state formulae for signal relations between synchronous paths and then in part (ii) we describe the Lorentz transformation formulae which relate natural coordinate systems with congruent time scales. (These results apply only to the special case of a collinear set).

Theorem 78 (see Figure 82a). *Let $\{S_t^x\}$ and $\{Q_{t'}^{x'}\}$ be natural coordinate systems in a collinear set Σ such that the paths $Q^{0'}$ and S^0 are synchronous and let r be the directed rapidity of $\{Q^{x'}\}$ with respect to $\{S^x\}$.*

(i) *Let S_u^0, $Q_{w'}^{0'}$, S_y^0 be events such that $F_{Q^{0'}S^0}^{+}(S_u^0) = Q_{w'}^{0'}$ and $F_{S^0Q^{0'}}^{-}(Q_{w'}^{0'}) = S_y^0$. Then*

$$w' = e^r u \quad \text{and} \quad y = e^r w' .$$

(ii) *If $Q_{t'}^{x'} = S_t^x$ then the coordinates are related by the "Lorentz transformation formulae":*

$$\begin{bmatrix} t' \\ x' \end{bmatrix} = \begin{bmatrix} \cosh r & -\sinh r \\ -\sinh r & \cosh r \end{bmatrix} \begin{bmatrix} t \\ x \end{bmatrix} \quad \text{and} \quad \begin{bmatrix} t \\ x \end{bmatrix} = \begin{bmatrix} \cosh r & \sinh r \\ \sinh r & \cosh r \end{bmatrix} \begin{bmatrix} t' \\ x' \end{bmatrix} .$$

(iii) *The path $Q^{x'_0}$ is the set of events with position–time coordinates $(x; t)$ related by*

$$x = x_0 + vt ,$$

where the directed velocity v is given by

$$v = \tanh r$$

and

$$x_0 = x'_0 \sqrt{1 - v^2} .$$

Remark The factor $\sqrt{1 - v^2}$ is known as the "Lorentz-Fitzgerald contraction factor".

Proof (i) By Theorem 72 and the definition of directed rapidity, $y = e^{2r} u$ and the paths $Q^{0'}$ and S^0 are synchronous which implies the equations of (i) (Figure 82a).

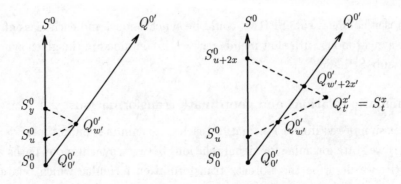

Figure 82a   Figure 82b

(ii) Next consider an event $Q_{t'}^{x'} = S_t^x$ (Figure 82b) which is on a right optical line containing events S_u^0 and $Q_{w'}^{0'}$, and which is on a left optical line which therefore contains events S_{u+2x}^0 and $Q_{w'+2x'}^{0'}$. Then

$$u = t - x \qquad \text{and} \qquad w' = t' - x' \tag{1}$$

so that

$$u + 2x = t + x \qquad \text{and} \qquad w' + 2x' = t' + x' \tag{2}$$

and by (i),

$$w' = e^r u \qquad \text{and} \qquad u + 2x = e^r(w' + 2x')$$

which, substituted into (1) and (2) respectively, gives

$$t' - x' = e^r(t - x) \qquad \text{and} \qquad t' + x' = e^{-r}(t + x)$$

which imply by addition and subtraction that

$$t' = (\cosh r)t - (\sinh r)x \tag{3}$$

$$x' = -(\sinh r)t + (\cosh r)x . \tag{4}$$

The second equation of (ii) can be obtained similarly, or by verifying that the square matrices are inverses of each other.

(iii) Now the path $Q^{x_0'}$ has x_0' constant, so equation (4) implies that

$$x = x_0'(\operatorname{sech} r) + (\tanh r)t .$$

Also by Theorem 71(ii),

$$x = x_0 + vt$$

and so we see that $v = \tanh r$ and therefore $\operatorname{sech} r = \sqrt{1 - v^2}$ so that $x_0 = x_0' \sqrt{1 - v^2}$.

q.e.d.

The equations of part (i) of the previous theorem illustrate the effect of rapidity on the times of transmission and reception of "light signals" between "synchronous observers". If this effect is composed for collinear paths, the composition of multiplicative factors leads to the addition of exponents and gives the same result as Theorem 73 of Section 8.1, namely that directed rapidity is an additive measure of speed.

The Lorentz transformation formulae (of part (ii)) describe a "pseudo–rotation" (see Theorem 66) between position–time coordinate systems. (It is instructive to compare and contrast these formulae with those for the rotation of orthogonal cartesian coordinates in the Euclidean plane.) The Lorentz transformation formulae may also be expressed in terms of velocity as

$$\begin{bmatrix} t' \\ x' \end{bmatrix} = \begin{bmatrix} \frac{1}{\sqrt{1-v^2}} & \frac{-v}{\sqrt{1-v^2}} \\ \frac{-v}{\sqrt{1-v^2}} & \frac{1}{\sqrt{1-v^2}} \end{bmatrix} \begin{bmatrix} t \\ x \end{bmatrix} \quad \text{and} \quad \begin{bmatrix} t \\ x \end{bmatrix} = \begin{bmatrix} \frac{1}{\sqrt{1-v^2}} & \frac{v}{\sqrt{1-v^2}} \\ \frac{v}{\sqrt{1-v^2}} & \frac{1}{\sqrt{1-v^2}} \end{bmatrix} \begin{bmatrix} t' \\ x' \end{bmatrix}.$$

The factor $\sqrt{1 - v^2}$ in the equation of (iii) is called the "Lorentz-Fitzgerald contraction factor" and expresses an apparent, although not real, "contraction of length" which is discussed in detail in many of the standard texts and is due to the asymmetric manner in which the lengths are specified. A second effect which is also due to an asymmetric manner of definition is the so-called "time dilation" or "relativistic slowing of moving clocks": the time coordinates t' and t for events $Q_{t'}^{0'} = S_t^x$ on the path $Q^{0'}$ are related, using the Lorentz transformation formulae, by the equation

$$t = \frac{1}{\sqrt{1 - v^2}} t'$$

which makes a single clock moving along the path $Q^{0'}$ appear to "run slowly" by comparison with several different clocks in the $\{S^x\}$ coordinate system. This effect applies to congruent clocks and is due to the asymmetric manner in which the time intervals are defined; that is, t' refers to the "local time" of a single clock whereas t refers to a time coordinate defined with respect to a set of "relatively stationary clocks".

9. Three–dimensional theorems

In this chapter we establish the isomorphism between $\mathcal{M}$ and the usual coordinate model M (which will be discussed in the following chapter). In Section 9.1 we show that each 3–SPRAY, with its paths considered as "points", is an "ordered" geometry in the sense described by Coxeter (1961, 1965). We show that a particular system of axioms given by Veblen (1904) is satisfied and this means that each 3–SPRAY is a convex subset of a three–dimensional projective space, with correspondences between paths and "points", and between collinear sets and "lines" defined according to the table of correspondences or "dictionary" prior to the statement of Theorem 80. This "dictionary" also includes correspondences between the usual projective geometry of three-dimensions and the line bundle model. In Section 9.2 we show that the paths of a given 3–SPRAY can be represented by a convex set of straight lines through the origin of $\mathbb{R}^4$ in the line bundle model and then in Section 9.3 we show that this set of lines is an ellipsoidal cone which means that a 3–SPRAY is a three–dimensional hyperbolic space. The purpose of Section 9.4 is to define coordinates on the set of events. The chapter concludes with Section 9.5 in which we establish the isomorphism between $\mathcal{M}$ and the usual coordinate model M in $\mathbb{R}^4$.

This enables us to describe paths as sets of events whose coordinates are specified by linear equations. Concepts of position–space, four–velocity, three–velocity, speed and relative rapidity are defined in relation to the particular coordinate system and their properties are briefly discussed. In particular it is shown that position–space is Euclidean.

9.1 Each 3–SPRAY is a 3–dimensional ordered geometry

We will show in Theorem 80 that each 3–SPRAY, with its paths considered as "points", may be considered to be a convex subset of a three–dimensional projective geometry. To do this we will show that a system of eleven axioms given by Veblen (1904) is satisfied with correspondences between the 3–SPRAY and the geometry defined by the "dictionary" which precedes the statement of the theorem. The axioms of Veblen are set out in Appendix 1 which also contains a brief description of the particular result we are using. So far we have established theorems which correspond to all, but one, of the required eleven axioms of Veblen. The remaining axiom corresponds to the following theorem.

Theorem 79 (Triangle Transversal)

Let A, B, C, D, E, be distinct paths which meet at an event x. If

(i) A, B, C *are not collinear*

(ii) $\langle B, C, D \rangle$

(iii) $\langle C, E, A \rangle$

then there is a path F such that $\langle A, F, B \rangle$ and $\langle D, E, F \rangle$.

Remarks 1. In order to prove this theorem we first establish Lemma 1 and Lemma 2.

2. Throughout the proof, the results of Theorem 71 apply to kinematic relations *within* the collinear set Σ.

Lemma 1 *Let Σ be a collinear set containing a path Q, let x be an event which does not belong to Σ. If $Q(x, \emptyset)$ has a first event a and a last event b, then no event of $\Sigma(x, \emptyset)$ is before a or after b.*

Proof We will show that there is no event of $\Sigma(x, \emptyset)$ after b by supposing the contrary and obtaining a contradiction. The proof is based on Theorem 13 which states that "the unreachable set is connected". Without loss of generality we consider an event c after b such that c is the last event of $bc(x, \emptyset)$ (Figure 83); then all events of the prolongation of bc after c are reachable (from x). (In this proof we shall abbreviate the expression "reachable from x" to "reachable".) Now the connectedness of the unreachable set (Theorem 13) implies that all events of $\mathcal{C}(a, b, c)$ are unreachable and, by the same theorem, all events after c and between the paths ac and bc are reachable (Region 1). Similarly all events before a and between the paths ac and ab are reachable (Region 2).

Since all events of Q after b are reachable, the same theorem (Th.13) implies that all events after b and to the left of Q are reachable (Region 3). Similarly all events before c and to the left of both Q and bc are reachable (Region 4).

Now let l_1 be the left optical line through $b_1 (:= b)$ and let r be the right optical line through c. Let $d_1 := l_1 \cap r$, let $Q^{(1)}$ be the parallel to Q through d_1 and let $b_2 := Q^{(1)} \cap bc$. The unreachable set $Q^{(1)}(x, \emptyset)$ contains more than one event (Axiom I5) and is connected, so the event d_1 is reachable and therefore the unreachable set $Q^{(1)}(x, \emptyset)$ precedes b_2 (and possibly contains b_2). Each event e after b_2 and to the left of $Q^{(1)}$ is after $Q^{(1)}(x, \emptyset)$, and is therefore reachable from x by Theorem 13 applied to a path which joins e to an event in $Q^{(1)}(x, \emptyset)$ and which

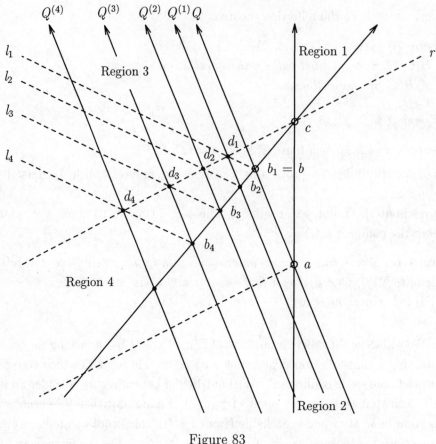

Figure 83

contains events of Region 4. Let l_2 be the left optical line through b_2: we have shown that the set of events after b_2 and between l_2 and Q is reachable from x. Similarly let $d_2 := l_2 \cap r$, let $Q^{(2)}$ be the parallel to Q through d_2 and let $b_3 := Q^{(2)} \cap bc$. Then the set of events after b_3 and between l_3 and Q is reachable.

An inductive argument based on sequences $b_n, l_n, d_n, Q^{(n)}$ together with an elementary argument about similar figures (based on the affine structure of Σ which was demonstrated in Theorem 71) shows that the entire set of events to the left of both Q and bc is reachable (Region α in Figure 84). Now take a path S parallel to ac and to the left of b_2. There is an event $e \in S(x, \emptyset)$ either (i) after b_2 or (ii) before b_2. In Case (i) there is a path eb_2 which meets ac in an event f before b_2. Again Theorem 13 implies that all events to the left of eb_2 and before b_2 are reachable (Region β). Now the parallel to ab through f lies within the union of Region α and

156

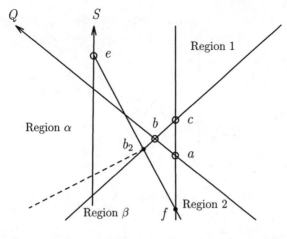

Figure 84

Region β and Region 2 so all its events are reachable: this contradicts Axiom I5. A similar contradiction is obtained in Case (ii).

We have now shown that there is no event of $\Sigma(x, \emptyset)$ after b. A similar argument shows that there is no event of $\Sigma(x, \emptyset)$ before a. The proof of Lemma 1 is now complete.

Lemma 2 *Let Σ be a collinear set which contains a kinematic triangle $\triangle abc$ and let x be an event which does not belong to Σ. If $\Sigma(x, \emptyset)$ meets one side of the triangle $\triangle abc$ internally, then it meets at least one other side.*

Proof Without loss of generality we consider the case where $\Sigma(x, \emptyset)$ meets $|ab|$ internally and where $a < b$. Then $ab(x, \emptyset)$ has a first event a' and a last event b' such that $a < a' < b' < b$ and, by the preceding lemma, $\Sigma(x, \emptyset)$ is contained within the set $\mathcal{W}$ of events which are not after b' and not before a'. It is clear that at least one of the two remaining sides $|ac|$ or $|bc|$ meets $\mathcal{W}$.

If $c \notin \mathcal{W}$ then either $ac \cap \mathcal{W} \subset (ac)$ or else $bc \cap \mathcal{W} \subset (bc)$, so either ac or else bc has its unreachable set located within the corresponding side $|ac|$ or $|bc|$, respectively.

In the other case where $c \in \mathcal{W}$ the events are ordered such that $a < c < b$ (Figure 85). Now if one side, say $|ac|$, does not meet $\Sigma(x, \emptyset)$, then $ac(x, \emptyset)$ must be on the prolongation of ac beyond (that is, after) c as well as being within $\mathcal{W}$. Furthermore, by the preceding lemma, $\Sigma(x, \emptyset)$ is contained within a set $\mathcal{W}^*$ of events

157

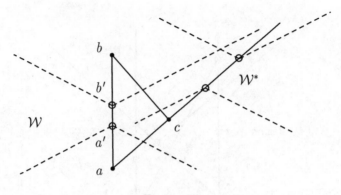

Figure 85

which are not after the last event of ac $(x, \emptyset)$ and not before the first event of ac $(x, \emptyset)$. Thus bc $(x, \emptyset)$ is contained in $\mathcal{W} \cap \mathcal{W}^*$ which is contained in the segment (bc).

A similar argument applies if $|bc|$ does not meet $\Sigma(x, \emptyset)$. This completes the proof of Lemma 2.

Proof (of the Triangle Transversal Theorem). Let $b \in B$ be an event other than the event x of coincidence. Theorem 4 implies the existence of an event $d \in D$ such that $[x, \ D(b, \emptyset), \ d]$ and, since $\langle B, C, D \rangle$, there is an event $c \in C$ such that $[bcd]$. Since unreachable sets are closed intervals (Theorem 16), there is an event $e' \in E$ such that $[x, \ e', \ E(c, \emptyset)]$ and such that, for any event $e \in E$ which satisfies $[xee']$, there are paths ce and de; moreover $[x, e', E(c, \emptyset)]$ and $\langle A, E, C \rangle$ imply (by Theorems 28 and 32) that ce meets A in an event a such that $[ce(x, \emptyset), \ a, \ e, \ c]$.

Let ce' meet A in the event a'. Since the unreachable set is a closed interval there is some event $a'' \in A$ with $[x \ a'' \ a']$ such that for all $a \in A$ which satisfy $[x \ a \ a'' \ (a')]$ there is a path ba. For any such event a, the Intermediate Path Theorem (Th.31) implies the existence of an event $e \in E$ such that $[c \ e \ a]$ and the Second Collinearity Theorem (Th.7) implies that $[x \ e \ e']$, so by the preceding paragraph there is a path de. Now by the First Collinearity Theorem (Th.3), the path de meets the segment (ab) in an event f such that $[a \ f \ b]$.

Now $[x, \ D(b, \emptyset), \ d]$ implies that $[bd(x, \emptyset), \ b, \ d]$ and we have already shown that $[ce(x, \emptyset), \ a, \ e, \ c]$. Thus all events between b and d as well as all events between a and c are reachable from x and so, by the preceding Lemma 2, all events of (ab) (and all events of (df)) are reachable from x. Therefore there is a path $F := xf$ such that $\langle D, E, F \rangle$ and $\langle A, F, B \rangle$. The proof of Theorem 79 is now complete. *q.e.d.*

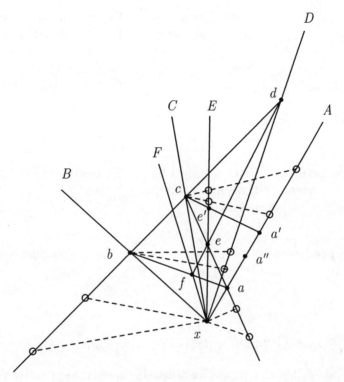

Figure 86   The paths A, B, C, D, E (with F) satisfy the triangle transversal theorem analogous to the axiom of Pasch

The next theorem (Th.80) shows that each 3–SPRAY is an "ordered geometry" in the sense described by Coxeter[1] (1961, 1965). After this theorem has been proved we will use familiar geometric terms with the correspondences as set out in the first two columns of the table on the following page. The theorem implies, by a result of Veblen (1904) (cited in Appendix 1 on p.214) that any 3–SPRAY, with its paths considered as "points", may be considered to be a convex subset of a three dimensional projective geometry. This convex subset may be represented with the line bundle model which is discussed in the following Section 9.2 with the correspondences as set out in the third column of the table.

Kinematic	Geometric	Standard Line Bundle Model
3–SPRAY	Three–dimensional ordered geometry	Convex bundle of straight lines $\mathcal{V}$ through O in $\mathbb{R}^4$
Paths $Q, R, S, \cdots \in 3SP$	Points $q, r, s, \ldots$	Ordinary lines $Q, R, S, \ldots$ through O
Collinear subset $CSP\langle Q, S \rangle$	Line qs	Plane $PL[Q, S]$
Sub–SPRAY spanned by three independent paths	Plane specified by three non–collinear points	Three–dimensional subspace through O

Table   Correspondences between a set of simultaneously coincident paths and an ordered geometry

Theorem 80 *Each 3–SPRAY is a three–dimensional ordered geometry*[1].

Proof   We will show that the axioms for ordered geometry, as given by Veblen (1904) (and stated in Appendix 1) are satisfied for a set of paths of a 3–SPRAY, where each path corresponds to a "point" and collinear sub–SPRAYS correspond to "lines". The properties of collinear sub–SPRAYS, as stated in Theorem 33 correspond to Veblen's Axioms II, III, IV, and VI, while the result of Theorem 52 implies that Veblen's Axiom V is satisfied.   Axioms I, VII, IX and X are all related to existence and dimension so they are satisfied for any given 3–SPRAY. The preceding Theorem 79 implies that Veblen's Triangle Transversal Axiom (Axiom VIII) is satisfied.   By Theorems 53 and 62, the set of paths of a collinear sub–SPRAY is order–isomorphic to the real numbers, so Veblen's Axiom XI (of Continuity) is satisfied.   Thus all eleven of Veblen's axioms for an ordered geometry[2] are satisfied. This completes the proof.   *q.e.d.*

9.2 The line bundle model of a 3–SPRAY

Projective space of three dimensions, P^3, can be described with respect to five suitably chosen points by a homogeneous coordinate system[3]. Once a particular homogeneous cordinate system has been specified, any point in P^3 is then represented by its homogeneous coordinates which are an equivalence class of ordered quadruples of reals, the equivalence being defined by proportionality. Thus a point in P^3 may be considered, with respect to the specified homogeneous coordinate system, as a line through the origin of $\mathbb{R}^4$ and a line in P^3 may be considered as a plane through the origin of $\mathbb{R}^4$. This model is called the *line bundle model* and is isomorphic to the homogenous coordinate model which is described in detail by Borsuk and Szmielew (1960) and Busemann and Kelly (1953). Borsuk and Szmielew (1960) use the model to demonstrate that P^3 is consistent and categorical.

The previous Theorem 80 implies, by the result of Veblen (1904) as stated in Appendix 1, that the set of paths of a 3–SPRAY is a convex subset of a three–dimensional projective space and can therefore be represented (to within an isomorphism) by a convex bundle of straight lines through the origin of $\mathbb{R}^4$.

In the next theorem we will be discussing properties of any given 3–SPRAY which will be denoted as $3SP$. In our line bundle model, the set of lines which correspond to paths of $3SP$ will be called *velocity space* and will be denoted as $\mathcal{V}$: these lines pass through the origin of $\mathbb{R}^4$. The correspondences between the line bundle model, the convex subset of projective three-space, and the paths of the 3–SPRAY are set out in the table which precedes Theorem 80. We will eventually develop and extend the familiar line bundle model of projective three-space in order to coordinatize Minkowski space–time, so we need to make some further definitions. Lines parallel to lines of $\mathcal{V}$ will be called *ordinary lines* and lines parallel to lines of the boundary $\partial\mathcal{V}$ of $\mathcal{V}$ will be called *limiting lines*. A plane in $\mathbb{R}^4$ spanned by two straight lines S and T has a *plane set of lines* $PL[\mathsf{S}, \mathsf{T}]$ and a *plane set of points* $pl[\mathsf{S}, \mathsf{T}]$, where we are adopting a convention that lines (resp. points) in $\mathbb{R}^4$ are denoted by upper (resp. lower) case sans serif symbols.

Theorem 81 (Homogeneous Coordinates for Paths of a 3–SPRAY)

(i) *Any given 3–SPRAY is an open convex subset of a three–dimensional projective geometry.*

(ii) *Any given 3–SPRAY can be represented to within an isomorphism by a convex set $\mathcal{V}$ of straight lines through the origin of $\mathbb{R}^4$, where each straight line corresponds to the (equivalence class of) homogeneous coordinates of a path in the 3–SPRAY. That is, there is a bijective identification mapping*

$$I : 3SP \longrightarrow \mathcal{V}$$
$$S \longmapsto \mathsf{S},$$

where corresponding lines are denoted with sans serif symbols.

(iii) *Isotropy mappings are collineations of the three–dimensional projective geometry.*

(iv) *An isotropy mapping θ induces an affinity θ^* of $\mathbb{R}^4$ such that, for each path $S \in 3SP$,*

$$I : \theta(S) \longmapsto \theta^*(\mathsf{S})$$

thus

$$I \circ \theta = \theta^* \circ I .$$

The induced mapping θ^* maps the set of ordinary lines bijectively onto itself and it maps the set of limiting lines bijectively onto itself.

(v) *A composition ψ of isotropy mappings (with possibly different invariant paths) induces an affine bijection $\psi^* : \mathbb{R}^4 \longrightarrow \mathbb{R}^4$ which bijectively maps the set of ordinary lines onto itself and bijectively maps the set of limiting lines onto itself.*

(vi) *For an isotropy mapping θ with invariant path Q, the induced mapping θ^* leaves the line Q invariant and maps the set of all lines parallel to Q onto itself.*

Remarks   1. For an isotropy mapping θ, the corresponding affinity θ^* is called an *induced isotropy mapping*.

2. To make the subsequent discussion more specific, we shall impose the further condition that an isotropy mapping θ with invariant path Q and corresponding invariant line Q has an induced isotropy mapping θ^* which leaves each point of the line Q invariant. This does not restrict the generality of the set of collineations (which correspond to those isotropy mappings which leave the path Q invariant),

since each straight line through the origin of $\mathbb{R}^4$ is mapped onto itself by a dilatation of $\mathbb{R}^4$.

Proof (i) By the previous theorem any given 3–SPRAY is a three–dimensional ordered geometry and can be embedded as a convex (three–dimensional) open subset in three–dimensional projective space[1]. The points of the convex subset are called "ordinary points".

(ii) Now three–dimensional projective space is categorical (that is, all models of it are isomorphic[4]). In the standard line bundle model of three–dimensional projective space[3], the "points" are represented as straight lines through the origin of $\mathbb{R}^4$ and the "lines" are represented as planes through the origin of $\mathbb{R}^4$: thus the set of "ordinary points" corresponds to a convex subset $\mathcal{V}$ of straight lines through the origin and, since the "ordinary points" are the paths of the given 3–SPRAY, these paths correspond to the convex subset $\mathcal{V}$ of straight lines through the origin. Any two paths specify a unique collinear set which, in turn, corresponds to a (two–dimensional) plane through the origin in $\mathbb{R}^4$. Thus there is a bijective correspondence

$$I: \qquad 3SP \longrightarrow \mathcal{V}$$
$$Q, R, S, \ldots \longmapsto \mathsf{Q}, \mathsf{R}, \mathsf{S}, \ldots$$

such that for paths $Q, R, S, \ldots$ there are lines $\mathsf{Q}, \mathsf{R}, \mathsf{S}, \ldots$ through the origin O of $\mathbb{R}^4$, and lines corresponding to paths are denoted by the corresponding sans serif symbols. For a collinear sub–SPRAY such as $CSP\langle S, T\rangle$, there corresponds the intersection of $\mathcal{V}$ with the plane $PL[\mathsf{S}, \mathsf{T}]$ (which contains both S and T and passes through the origin O of $\mathbb{R}^4$). This completes the proof of (ii).

(iii) Isotropy mappings preserve the incidence relation (Axiom S) so the Intermediate Path Theorem (Th.31) implies that isotropy mappings preserve the relation of betweenness $\langle \cdots \rangle$ for paths. Thus isotropy mappings are collineations on the subset of "ordinary points". This subset is open, three–dimensional and convex. Therefore isotropy mappings induce collineations on the entire projective space. This completes the proof of (iii).

Before we establish the result (iv) we briefly review some properties of three–dimensional projective geometry. In the usual model[3] the homogeneous coordinates

$$\left\{ \lambda(y_0, y_1, y_2, y_3) : \ (y_0, y_1, y_2, y_3) \neq (0,0,0,0), \ \lambda \neq 0, \ \lambda \in \mathbb{R}^4 \right\}$$

correspond to a line through the origin $O(0,0,0,0)$ in $\mathbb{R}^4$. Consequently we can use analytic techniques to describe three–dimensional projective geometry. Any collineation in P^3 can be described by a linear mapping

$$\theta^* : \quad x_i \longmapsto x_i' = \sum_{j=0}^{3} a_{ij} x_j \qquad (\det a_{ij} \neq 0)$$

which is clearly an affinity of $\mathbb{R}^4$. Now for any path $S \in 3SP$,

$$I : \quad \theta(S) \longrightarrow \theta^*(\mathsf{S}) = \theta^*(I(S))$$

so

$$I(\theta(S)) = \theta^*(I(S))$$

whence

$$I \circ \theta = \theta^* \circ I \quad .$$

With respect to an isotropy mapping θ, the (set of paths of the) given 3–SPRAY is mapped bijectively to itself, so the set of lines of $\mathcal{V}$ is mapped bijectively onto itself, hence the set of ordinary lines is mapped bijectively onto itself and so is the set of limiting lines (which are parallel to lines of the boundary $\partial \mathcal{V}$ of $\mathcal{V}$): this completes the proof of (iv).

To establish the result (v) we simply observe that the composition of two affine bijections is an affine bijection and, by part (iv), $\mathcal{V}$ and $\partial \mathcal{V}$ are invariant under each affinity so they are invariant with respect to the composition.

The result (vi) is an immediate consequence of (iv). $\qquad\qquad$ q.e.d.

9.3 Each 3–SPRAY is a 3-dimensional hyperbolic space

In the next theorem, we will show that each 3–SPRAY is a three–dimensional hyperbolic space using a result of Busemann (1955)[5]. The theorem will be established for the convex set of ordinary lines $\mathcal{V}$ which, by the preceding theorem, represents the paths of a 3–SPRAY. We will change the coordinatization of $\mathbb{R}^4$ in order to describe $\mathcal{V}$ in the simplest possible way. The space $\mathbb{R}^4$ is a four-dimensional affine space A^4 which will be equipped with a suitable system of coordinates: then its set of points will be denoted as E, individual points are denoted by lower case sans serif symbols a, b, c, d, e, x, .. with corresponding coordinates $a_i, b_i, c_i, d_i, e_i, x_i, ..$ in the usual mathematical italic font. Each straight line has a parametric equation of the form

$$x_i = x_{i(initial)} + \lambda w_i \qquad (i = 0, 1, 2, 3)$$

where the four-component direction vector w_i is called a *4–velocity vector*. We will use the index conventions that italic letters $i, j, k, \ldots$ range over the integers $0, 1, 2, 3$

while Greek letters $\alpha, \beta, \gamma, \ldots$ range over the integers $1, 2, 3$ and repeated subscripts imply a sum in accordance with the Einstein summation convention.

Theorem 82 (Each 3–SPRAY is a three–dimensional hyperbolic space)
Let A^4 be a four-dimensional affine space and let $\mathcal{V}$ be a convex set of lines through some given point O of A^4 with Q being a line contained in the interior of $\mathcal{V}$. If, for any two lines $R', S' \in \partial\mathcal{V}$, there is an affinity θ^* such that

$$\theta^* : \quad \mathcal{V} \longrightarrow \mathcal{V}, \quad R' \longrightarrow S'$$

and all points of Q are invariant, then $\mathcal{V}$ is an ellipsoidal cone. That is, the points of A^4 can be coordinatized such that each line has a parametric equation of the form

$$x_j = x_{j(initial)} + \mu w_j, \qquad \mu \in \mathbb{R}^4$$

where $w_j = (w_0, w_1, w_2, w_3)$ is a constant direction vector and
(i) *ordinary lines have direction vectors such that*

$$w_0^2 - w_1^2 - w_2^2 - w_3^2 > 0$$

(ii) *limiting lines have direction vectors such that*

$$w_0^2 - w_1^2 - w_2^2 - w_3^2 = 0 \qquad and$$

(iii)

$$\theta^* : \quad x_i \longrightarrow x_i' = \sum_{j=0}^{3} a_{ij} x_j \qquad (det\ a_{ij} \neq 0)$$

where

$$a_{ij} = \begin{bmatrix} 1 & 0 \\ 0 & a_{\alpha\beta} \end{bmatrix}$$

and the sub-matrix $a_{\alpha\beta}$ is orthogonal.

Remarks   1. The existence of the affinity θ^* which maps R' onto S' is implied by the Axiom of Isotropy (Axiom S).

2. The theorem implies that $\mathcal{V}$ is an ellipsoidal cone whose lines can be regarded as the "points" of a three–dimensional hyperbolic geometry.

Proof   In the affine space A^4, there is an affine parameterisation of the points of the line Q such that Q_0 is the vertex of (the set of lines of) $\mathcal{V}$.

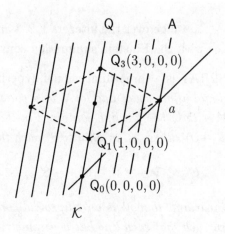

Figure 87   The set of parallels $\mathcal{K}$

(a) *Definition of the set of parallels $\mathcal{K}$.*   The set of straight lines through Q_0 is a three–dimensional projective space with each line being a "point" of the space.   Any plane through Q_0 contains a set of lines through Q_0 which are the "points" of a "straight line"of the projective space.   Now, considering A^4, in each plane which contains Q, there are two limiting lines through Q_1, and two limiting lines through Q_3 (Figure 87). A pair of non–parallel limiting lines through these points meets in some point a which can be joined to Q_0 by an ordinary line (since $\mathcal{V}$ is convex): moreover there is a line A parallel to Q such that $a \in A$. Thus in each plane which contains Q there are two such parallels, one on either side of Q: each of these two parallels corresponds to the intersection of two limiting lines — one through Q_1 and the other through Q_3. Any two ordinary lines which pass (respectively) through Q_1 and Q_3 and which are parallel (respectively) to lines of $\mathcal{V}$ on opposite sides of Q, meet in a point through which there is a parallel to Q: since $\mathcal{V}$ is convex this parallel is between the previously mentioned pair of parallels generated by the intersection of limiting lines. In each plane the set of all such parallels (including the pair defined by the limiting lines) is connected, contains Q and is bounded by the pair of parallels which were defined by the limiting lines This procedure applies to each plane through Q: the set of all such parallels is denoted by $\mathcal{K}$.

(b) *The set of parallels $\mathcal{K}$ is an ellipsoid.*   Since $\mathcal{V}$ is convex, the set of half–lines (corresponding to the ordinary and limiting lines of $\overline{\mathcal{V}}$) is separated into two components by any of the supporting "planes" (hyperplanes). Hence the convexity

166

of (both components) implies the convexity of $\mathcal{K}$. Furthermore, $\overline{\mathcal{V}}$ is convex and each "line" (plane) containing Q has two distinct boundary "points" (limiting lines), so $\overline{\mathcal{V}}$ is bounded and hence the set $\mathcal{K}$ of parallels is bounded.

Now by the previous theorem (Th.81(iv)), the mapping θ^* is an affinity and therefore induces a bijection Ψ of the set of parallels (to Q) onto itself. Furthermore

$$\theta^* : \quad Q_1 \longrightarrow Q_1, \quad Q_3 \longrightarrow Q_3, \quad \mathcal{V} \longrightarrow \mathcal{V}$$

and since the set of ordinary lines is mapped bijectively onto itself, the set of limiting lines is also mapped bijectively onto itself and therefore θ^* maps $\mathcal{K}$ bijectively onto itself. Thus the preconditions of the characterization of ellipsoids given by Busemann[5] are satisfied: $\mathcal{K}$ is a bounded convex body in a three–dimensional affine space (of parallels to Q) which contains Q as an interior "point" (line) and for any two "points" (parallels) $A, B \in \partial\mathcal{K}$, an affinity exists that leaves Q fixed, maps $\partial\mathcal{K}$ bijectively onto itself and A to B. It follows by the result of Busemann that the set of parallels $\mathcal{K}$, interpreted as "points", form a three–dimensional ellipsoid.

(c) *Coordinatization of the Set of Parallels.* Since $\mathcal{K}$ is an ellipsoid, the three–dimensional affine space of parallels (to Q) can be coordinatized with a set of Y_α–coordinates ($\alpha = 1,2,3$) such that the set of all parallels of $\mathcal{K}$ satisfy the inequality

$$y_1^2 + y_2^2 + y_3^2 \leq 1 \tag{1}$$

where the equality corresponds to the parallels of the boundary $\partial\mathcal{K}$.

(d) *Coordinatization of the Set of Points.* The points of A^4 can now be coordinatized with a set of X_i–coordinates ($i = 0, 1, 2, 3$) in the following way: each point lies on some parallel with coordinates (y_1, y_2, y_3) so we define the coordinates X_α ($\alpha = 1, 2, 3$) for the point to be the same as for the parallel. The line Q has coordinates $(y_1, y_2, y_3) = (0, 0, 0)$ and its points have already been indexed with an affine parameter, so for each point $Q_t \in Q$ we define $x_0 = t$ and then the coordinates of Q_t are $(t, 0, 0, 0)$.

On each plane through Q there are two bounding parallels in $\partial\mathcal{K}$ such that

$$y_1^2 + y_2^2 + y_3^2 = 1 \ . \tag{2}$$

We consider one such parallel A and (as discussed in the second paragraph of this proof) there are limiting lines which pass through the points $(1, 0, 0, 0)$ and $(3, 0, 0, 0)$ and which meet at a point $a \in A$: we specify that the x_i–coordinates of a are

$(2, y_1, y_2, y_3)$ and for the limiting line which passes through $(1,0,0,0)$ and through $a(2, y_1, y_2, y_3)$ we specify the point coordinates x_i with the (four) equations

$$(x_0, x_1, x_2, x_3) = (1,0,0,0) + \mu(1, y_1, y_2, y_3) \ , \qquad \mu \in \mathbb{R} \ . \tag{3a}$$

These equations specify the x_0–coordinate at one point on each parallel to Q in the plane which contains Q and A. Now take a limiting line through the point $(3,0,0,0)$ parallel to the previous limiting line. Let the point at which this line meets A have the coordinates $(4, y_1, y_2, y_3)$. Then this line has the parametric equations

$$(x_0, x_1, x_2, x_3) = (3,0,0,0) + \mu(1, y_1, y_2, y_3) \ , \qquad \mu \in \mathbb{R} \ . \tag{3b}$$

These equations specify an x_0–coordinate on a second point on each parallel to Q in the plane spanned by Q and A. Thus each parallel has been equipped with an affine parameter and hence the plane has been given an affine coordinatization.

Since each plane through Q passes through some bounding parallel of $\partial \mathcal{K}$ (satisfying equation (2)), it follows that the entire affine space A^4 is equipped with an affine coordinatization: *the set of all points coordinatized in this way will be denoted by E.*

(e) *Parametric Equations of Ordinary Lines and Limiting Lines.* Any line has an equation of the form

$$x_i = x_{i(initial)} + \mu w_i \ , \qquad \mu \in \mathbb{R} \tag{4}$$

(for i=0,1,2,3) where μ is a line parameter and w_i is a constant direction vector. Equations (2) and (3) imply that for limiting lines

$$w_0^2 - w_1^2 - w_2^2 - w_3^2 = 0 \tag{5a}$$

hence for ordinary lines

$$w_0^2 - w_1^2 - w_2^2 - w_3^2 > 0 \tag{5b}$$

which completes the proof of parts (i) and (ii).

To establish the result (iii) we observe firstly that θ^* bijectively maps the set of parallels $\partial \mathcal{K}$ (which satisfy the equality (2)) onto itself and secondly that (by definition) θ^* leaves the points of Q invariant: thus the x_0–coordinate is invariant with respect to θ^*. The induced mapping on the Y_α–coordinates preserves the equality (2) so it is an orthogonal mapping and since the triples (x_1, x_2, x_3) and (y_1, y_2, y_3) are equal it follows that θ^* maps the $X_1 X_2 X_3$–subspace orthogonally. Hence the associated matrix a_{ij} is partitioned as in (iii) and the sub-matrix $a_{\alpha\beta}$ is orthogonal.

q.e.d.

Equipped with these coordinates, the set of points will be denoted as E, the set of ordinary lines will be denoted as $\mathcal{L}$ and the relation of betweenness $[\cdots]$ will be defined in the obvious way on each ordinary line. We define the *standard model* to be

$$M := \langle E, \mathcal{L}, [\cdots] \rangle$$

which is related to the usual affine space A^4, coordinatized as in the statement of Theorem 82, but with the subset $\mathcal{L}$ of lines, rather than the usual set of straight lines of A^4. The purpose of the next two sections is to establish an isomorphism between $\mathcal{M}$ and M.

9.4 Coordinatization theorem

The Axioms of Existence and Dimension (Axioms I1 and I3) ensure the existence of a path Q which has a 3–SPRAY at the event Q_0. We will specify a system of coordinates for the (subset of) events which belong to those collinear sets which contain the path Q. This coordinate system is defined in such a way that collinear sets containing Q correspond to planes containing the corresponding line Q. In the next theorem we are able to establish the isomorphism between these particular collinear sets and the corresponding planes. The important isomorphism between $\mathcal{M}$ and M will be established in Theorem 86 of the following Section 9.5. Until then, we shall work with a lesser concept which will be called a "path isomorphism": a subset $\mathcal{D} \subseteq \mathcal{E}$ is *path isomorphic* to a subset $D \subseteq E$ and we write $\mathcal{D} \simeq D$ if there is a bijection from $\mathcal{D}$ to D such that for events $b, c \in \mathcal{D}$ and the corresponding points $\mathsf{b}, \mathsf{c} \in D$,

$$\exists\, bc \in \mathcal{P} \quad \Longleftrightarrow \quad \exists\, \mathsf{bc} = (b_0, b_1, b_2, b_3)(c_0, c_1, c_2, c_3) \in \mathcal{L} .$$

Note that the concept of path isomorphism makes no reference whatsoever to relations of betweenness between events of $\mathcal{M}$ or points of M. We denote path isomorphism by the symbol $\simeq$ and use the symbol $\cong$ to denote the stronger property of *affine–isomorphism*.

In the next theorem, we will extend the identification mapping from

$$I : 3SPR[Q_0] \longrightarrow \mathcal{V}$$

to an identification mapping which applies from the union of all collinear sets of events which contain Q to the points of E, namely

$$i : \bigcup_{R \in 3SPR[Q_0]} col[Q, R] \longrightarrow E .$$

Let Q be a path with a natural time–scale such that there is a 3–SPRAY with common event, $3SPR[Q_0]$ and let there be an identification mapping

$$I : 3SPR[Q_0] \longrightarrow \mathcal{V}$$

where $\mathcal{V}$ is a set of lines through the origin of the affine space M as in the statement of Theorem 81. Define the *identification mapping*

$$i_Q : Q \longrightarrow \mathbf{Q}$$

between events of the path Q and points of the line $\mathbf{Q}$ such that

$$i_Q : Q_t \longrightarrow (t, 0, 0, 0) .$$

Theorem 83 (Coordinatization)
For each path $R \in 3SP[Q_0]$ there is a bijection, between the events of $col[Q, R]$ and the points of the plane $pl[\mathbf{Q}, \mathbf{R}]$, which induces a bijection between the paths of $COL[Q, R]$ and the ordinary lines of $PL[\mathbf{Q}, \mathbf{R}]$: for each event of $col[Q, R]$ with position–time coordinates $(x; t)$, the corresponding point in $pl[\mathbf{Q}, \mathbf{R}]$ has coordinates (x'_0, x'_1, x'_2, x'_3) such that

$$t = x'_0 \qquad and \qquad x = \sqrt{(x'_1)^2 + (x'_x)^2 + (x'_3)^2} .$$

Thus the identification mappings $i_Q : Q \longrightarrow \mathbf{Q}$ and $I : 3SP[Q_0] \longrightarrow \mathcal{V}$ imply the existence of a bijective coordinate mapping

$$i : \bigcup_{R \in 3SP[Q_0]} col[Q, R] \longrightarrow E$$

$$e \longmapsto (e_0, e_1, e_2, e_3)$$

such that, for each $R \in 3SP[Q_0]$,

(i) $i: R \longrightarrow I(R) = R$

(ii) $col[Q, R] \cong pl[Q, R]$ *and*

(iii) *for any isotropy mapping θ with invariant path Q,*

$$\theta^* \circ i = i \circ \theta .$$

Remark This theorem establishes the affine isomorphism (ii) but does not establish an affine isomorphism (or even a path isomorphism) between $\bigcup_{R \in 3SP[Q_0]} col[Q, R]$ and E. This important isomorphism will be established in the subsequent Theorem 86.

Proof (i,ii) Theorem 71(ii) implies that for any collinear set, the space of position–time coordinates is $\mathbb{R}^2$, the structure of paths and optical lines is affine and the equivalence classes of position–time coordinates for paths which pass through the origin $O(0; 0)$ are homogeneous coordinates. Each path has $|v| < 1$, so any event with position–time coordinates $(x; t)$ lies on a path through the origin if and only if

$$t^2 - x^2 > 0 . \tag{1}$$

The $X_0 X_1$–plane in M (which has $x_2 = 0$ and $x_3 = 0$) is also affine and isomorphic to $\mathbb{R}^2$. The straight lines through the origin correspond to paths if and only if they lie within V (see Theorem 81); that is, if and only if

$$x_0^2 - x_1^2 > 0 . \tag{2}$$

The set of all ordinary lines in this plane is therefore isomorphic to the set of all paths in a collinear set with the correspondences

$$x_0 = t \quad \text{and} \quad x_1 = x . \tag{3a}$$

Each plane through Q has a set of lines parallel to lines of V and this set of lines is affine–isomorphic to the set in the $X_0 X_1$–plane. Furthermore Theorem 82 (iii) implies that for each plane there is a matrix a_{ij} with $a_{00} = 1, a_{\alpha 0} = a_{0\alpha} = 0$, and $a_{\alpha\beta}$ orthogonal which maps the plane and its set of lines bijectively onto the $X_0 X_1$–plane with the correspondences

$$x_0' = x_0 \quad \text{and} \quad \sqrt{(x_1')^2 + (x_2')^2 + (x_3')^2} = x_1 . \tag{3b}$$

Combining equations (3a) and (3b) we see that each plane through Q has a set of ordinary lines which is isomorphic to the set of paths in a collinear set and, in particular, to the collinear set of paths $COL[Q, R]$ with the correspondences

$$x_0' = t \quad \text{and} \quad \sqrt{(x_1')^2 + (x_2')^2 + (x_3')^2} = x . \tag{3}$$

between coordinates of M and position–time coordinates of the collinear set.

Thus we extend the identification mapping i_Q to the mapping

$$i: \bigcup_{R \in 3SPR[Q_0]} col[Q, R] \longrightarrow E$$

which is defined such that, for each $R \in 3SP[Q_0]$,

$$i: \qquad R \longrightarrow \mathsf{R} = I(R)$$
$$col[Q, R] \longrightarrow pl[\mathsf{Q}, \mathsf{R}] = pl[I(Q), I(R)]$$

and then for each event $e \in col[Q, R]$ with position–time coordinates $(x; t)$,

$$i: \; e \longmapsto x'_i$$

such that

$$t = x'_0 \qquad \text{and} \qquad x = \sqrt{\left(x'_1\right)^2 + \left(x'_2\right)^2 + \left(x'_3\right)^2} \quad . \tag{4}$$

Then

$$col[Q, R] \; \cong \; pl[\mathsf{Q}, \mathsf{R}] \; .$$

(iii) Theorem 81 implies that for each $S \in SPR[Q_0]$ and for each isotropy mapping θ with invariant path Q,

$$\theta^*(i(S)) = i(\theta(S)) \; .$$

Since, in particular, this is true for all $S \in CSP\langle Q, R \rangle$ and since for all $Q_t \in Q$,

$$\theta^*(i(Q_t)) = i(\theta(Q_t)) = (t, 0, 0, 0)$$

it follows that for each event $e \in csp\langle Q, R \rangle$,

$$\theta^*(i(e)) = i(\theta(e))$$

and hence, given the affine structure of both $col[Q, R]$ and $pl[\mathsf{Q}, \mathsf{R}]$, it follows that for each event $e \in col[Q, R]$,

$$\theta^* \circ i = i \circ \theta^* \; .$$

$$q.e.d.$$

9.5 Isomorphism with the standard model

In this section we prove the important Isomorphism Theorem (Th.86). Before doing this, we obtain a preliminary result (Theorem 84) which establishes a path isomorphism from a subset of events which occur "after" the event Q_0 in the sense to be defined below.

In Theorem 85 we will extend this result to apply to an event Q_{-n} (for $n = 0, 1, 2, \ldots$). Since the index n is arbitrary, we will be able to extend the result to apply to the entire set of events and thereby establish the Isomorphism Theorem.

On each collinear set which contains the path Q, there is a temporal order relation specified by the natural time scale of events of Q. That is, for each event $Q_{-n}(n = 0, 1, 2, \ldots)$ and each path $R \in SPR[Q_{-n}]$ we can specify the *temporal order relation* $<$ on $col[Q, R]$, in accordance with the natural time scale on Q, and then we can define the *future sub-spray*

$$spr^+[Q_{-n}] := \{z : \ Q(z, \emptyset) > Q_{-n}, \ z \in spr[Q_{-n}]\} \ .$$

Similarly we define

$$E_{-n} := \{z_j : \ z_0 > -n \ \text{ and } \ (-n, 0, 0, 0)(z_0, z_1, z_2, z_3) \in \mathcal{L}, \ z_j \in E\} \ .$$

Theorem 84 (Path Isomorphism)

$$spr^+[Q_0] \simeq E_0 \ .$$

Proof (i) We first show that the identification mapping i induces a bijection between the two sets. By Theorem 83,

$$
\begin{aligned}
& Q(z, \emptyset) > Q_0 \quad \text{and} \quad z \in spr[Q_0] \\
\Longleftrightarrow \ & Q(z_i, \emptyset) > 0 \quad \text{and} \quad (0, 0, 0, 0)(z_0, z_1, z_2, z_3) \in \mathcal{L} \\
\Longleftrightarrow \ & z_0 > 0 \quad \text{and} \quad (0, 0, 0, 0)(z_0, z_1, z_2, z_3) \in \mathcal{L} \ . \quad (1)
\end{aligned}
$$

(ii) Next we will show that, for any two events $b, c \in spr^+[Q_0]$ with corresponding points $\mathsf{b, c} \in E$

$$\exists \ \mathsf{bc} = (b_0, b_1, b_2, b_3)(c_0, c_1, c_2, c_3) \in \mathcal{L} \quad \Longleftrightarrow \quad \exists \ bc \in \mathcal{P} \ .$$

This will establish that a path isomorphism exists between $spr^+[Q_0]$ and E_0.

Since $b, c \in spr^+[Q_0]$ there are paths $S := Q_0 b$ and $T := Q_0 c$. By Theorem 66 there is a composition ψ of two isotropy mappings such that $\psi : S \longrightarrow Q$. By the Axiom of Isotropy (Axiom S) and Theorem 21, the mapping ψ is a bijection of $\mathcal{E}$ to $\mathcal{E}$ and it induces a bijection of $\mathcal{P}$ to $\mathcal{P}$. Thus, if we let $U := \psi(T)$, we see that ψ is a path isomorphism of $col[S, T]$ to $col[Q, U]$ and so

$$\exists \, bc \in \mathcal{P} \quad \Longleftrightarrow \quad \exists \, \psi(b)\psi(c) \in \mathcal{P} . \tag{2}$$

Since $\psi(b), \psi(c) \in col[Q, U]$, Theorem 83(ii) applies so

$$\exists \, \psi(b)\psi(c) \in \mathcal{P} \quad \Longleftrightarrow \quad \exists \, \psi^*(b_j)\psi^*(c_j) \in \mathcal{L} \tag{3}$$

Since ψ is a composition of isotropy mappings, Theorem 81(v) implies that

$$\exists \, \psi^*(b_j)\psi^*(c_j) \in \mathcal{L} \quad \Longleftrightarrow \quad \exists \, (b_0, b_1, b_2, b_3)(c_0, c_1, c_2, c_3) \in \mathcal{L} . \tag{4}$$

Finally, the combination of $(2), (3), (4)$ completes the proof of (ii). *q.e.d.*

Theorem 85 *For each non-negative integer n,*

$$spr^+[Q_{-n}] \simeq E_{-n}$$

Proof (i) We first show that $spr^+[Q_{-n}] \subseteq \bigcup_{R \in 3SP[Q_0]} col[Q, R]$. For any event $e \in spr^+[Q_{-n}]$ the unreachable set $Q(e, \emptyset)$ is after Q_{-n} and Theorem 16 implies that there is some path R in $SPR[Q_0] \backslash \{Q\}$ which meets the path eQ_{-n} at some event. Thus $e \in col[Q, R]$ and $SPR[Q_{-n}]$ is a SPRAY whose dimension is no greater than that of $SPR[Q_0]$.

(ii) An argument similar to that of (i) above, shows that the dimension of $SPR[Q_0]$ is not greater than the dimension of $SPR[Q_{-n}]$. Thus $SPR[Q_{-n}]$ and $SPR[Q_0]$ both have dimension 3.

(iii) Now apply two coordinate transformations. For each collinear set which contains Q, transform the position–time coordinates from $(x; t)$ to $(x'; t') = (x, t + n)$ and for the points of E, transform the coordinates from (x_0, x_1, x_2, x_3) to $(x'_0, x'_1, x'_2, x'_3) = (x_0 + n, x_1, x_2, x_3)$. Then the same argument as for the preceding Theorem 84 shows that, for the primed coordinate systems,

$$spr^+[Q_{0'}] \simeq E_{0'} .$$

But $spr^+[Q_{0'}] = spr^+[Q_{-n}]$ and $E_{0'} = E_{-n}$, so the proof is complete. *q.e.d.*

In the final theorem of this chapter, we show that Minkowski space–time $\mathcal{M}$ and the standard model M are isomorphic and that automorphisms of $\mathcal{M}$ correspond to affinities of M.

Theorem 86 (Isomorphism)

(i)

$$\mathcal{M} \cong M .$$

That is

$$\langle \mathcal{E}, \mathcal{P}, [\cdots] \rangle \cong \langle E, \mathcal{L}, [...] \rangle$$

where the symbol $\cong$ denotes the usual concept of isomorphism.

(ii) *The automorphisms of $\mathcal{M}$ correspond to affinities of M.*

Remarks   1. In this theorem we establish isomorphism rather than the weaker concept of path isomorphism defined in Section 9.4.

2. We have already observed that induced isotropy mappings are affinities. The result (ii) applies to all automorphisms of $\mathcal{M}$.

Proof   We will first show that $\mathcal{E} = \bigcup_n spr^+[Q_{-n}]$.

Take any event $e \in \mathcal{E} \setminus spr^+[Q_0]$. The Axiom of Existence (Axiom I2) implies that there are paths S, T and an event a such that

$$e \in S , \quad a \in S \cap T , \quad Q_0 \in T .$$

By Theorems 4 and 6 there is an event $b \in S$ and a path bQ_0 such that $[aeb]$. Theorem 34 implies that one of the two events a, b belongs to $spr^+[Q_0]$ while the other event does not: without loss of generality, we shall regard a as the event which belongs to $spr^+[Q_0]$. Then there is an event Q_{-n} such that $[Q_0, Q(b, \emptyset), Q_{-n}]$ and hence $b \in spr^+[Q_{-n}]$. Now

$$n' > n \implies spr^+[Q_{-n'}] \supseteq spr^+[Q_{-n}]$$

so $a, b \in spr^+[Q_{-n}]$ and hence $e \in spr^+[Q_{-n}]$. But e is an arbitrary event of $\mathcal{E} \setminus spr^+[Q_0]$ and hence $\mathcal{E} \subseteq \bigcup_n spr^+[Q_{-n}]$. The opposite containment is a consequence of the definition of $spr^+[Q_{-n}]$, whence $\mathcal{E} = \bigcup_n spr^+[Q_{-n}]$.

(i) By the previous theorem,

$$spr^+[Q_{-n}] \simeq E_{-n}$$

and, for $n' > n$,

$$spr^+[Q_{-n'}] \supseteq spr^+[Q_{-n}] \qquad \text{and} \qquad E_{-n'} \supseteq E_{-n}$$

whence

$$\mathcal{E} \simeq E.$$

Now that the path isomorphism $\mathcal{E} \simeq E$ has been established, the identification mapping between events of the path Q and points of the line $\mathbf{Q}$ (see the preceding Coordinatization Theorem 83) implies that the betweenness relation $[\cdots]$ on (the paths of) $\mathcal{M}$ corresponds to the obvious betweenness relation on (the ordinary lines of) M. Thus $\mathcal{M} \cong M$ in the sense that

$$\langle \mathcal{E}, \mathcal{P}, [\cdots] \rangle \cong \langle E, \mathcal{L}, [\cdots] \rangle$$

where $\cong$ denotes the usual concept of isomorphism.

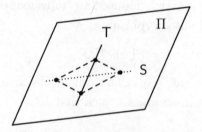

Figure 88

(ii) In the affine space M there are ordinary lines and limiting lines and also other straight lines which, in the literature, are called "*spacelike lines*": these spacelike lines are parallel to lines which belong to the exterior of $\mathcal{V}$. Since $\mathcal{V}$ is a convex cone, each line of either of the three types belongs to some plane which contains an ordinary line.

In each plane Π which contains an ordinary line there are two parallel classes of limiting lines which can be used to specify the usual concept of "equality of intervals" on classes of parallel ordinary lines (as in the proof of Theorem 63). The same concept can then be extended to limiting and spacelike lines in the plane. For any given limiting line in the plane there is a distinct class of parallel limiting lines which project equal intervals from any ordinary line of the plane onto the given limiting line. Similarly given any spacelike line S in the plane and any two distinct

176

events of S, there are two pairs of limiting lines which pass through the events and which meet at two distinct events which lie on some ordinary line T in the plane (as in Figure 88). Properties of similarity imply that each event on the spacelike line S can be specified by a pair of events of T which are at equal affine intervals from the intersection of T and S: furthermore, the interval measure on T can be projected by either one of the classes of limiting lines to define an affine measure of intervals on S. Thus the affine structure of the plane and hence of M is specified by the set of ordinary and limiting lines which, in turn, are specified by the set of ordinary lines.

Consequently each bijection of the set of ordinary lines preserves the affine structure of M and therefore is an affinity. $\qquad$ *q.e.d.*

The Isomorphism Theorem implies that each event corresponds to a point in 4-dimensional affine space. Now that we have established the isomorphism, it is no longer necessary to distinguish between $\mathcal{M}$ and M and, accordingly, we will have no further need to use sans serif symbols. Paths correspond to ordinary lines which are conventionally known as *timelike lines* and limiting lines are conventionally known as *lightlike lines*. Lines other than timelike and lightlike lines are known as *spacelike lines*. Vectors are similarly described as being (respectively) *timelike-*, *lightlike-*, or *spacelike-vectors*. In particular the direction vectors of timelike and lightlike lines (as described in Theorem 82) are known as *4-velocity vectors* or simply as *4-velocities*.

A. Space–time coordinates relative to Q   Space–time can now be described with reference to the coordinate system specified in the Isomorphism Theorem. Each event $x \in \mathcal{E}$ can be represented by its coordinate quadruple $x_i \in E$. The coordinate system, together with the parametric equations of timelike and lightlike lines, has been described in Theorems 82 and 83. It is conventional, when referring to one particular coordinate system, to use the word "event" to refer both to the actual event and to its coordinate quadruple. (In Section 10.5 we shall consider other coordinate systems and transformations between coordinate systems. At that stage it will become necessary to clearly distinguish between events and their different coordinate representations).

The set of paths parallel to Q is called the *position space (relative to Q)*. Each path of the position space has 4-velocity $(1, 0, 0, 0)$ and is specified by a coordinate triple (x_1, x_2, x_3) called its *position–space coordinates*. Each event of $\mathcal{M}$ is specified by four coordinates (x_0, x_1, x_2, x_3): the 0-coordinate is called the *time coordinate* and

177

the remaining coordinate triple is called its *(position) space coordinates*; together the four coordinates are called *space–time coordinates*.

B. Euclidean geometry of position space   Given any two paths S and T in position space relative to Q, there are two families of lightlike lines between them and Theorem 82(ii) implies that for any lightlike line between this pair of paths, the difference in the time coordinates between events $s_i \in S$ and $t_i \in T$ is given by the *distance function*

$$d_{TS} := \left((t_1 - s_1)^2 + (t_2 - s_2)^2 + (t_3 - s_3)^2\right)^{1/2} .$$

This is a Euclidean metric function on position space, so we conclude that *position space (relative to Q) is Euclidean.*

C. Measures of velocity, speed and rapidity   Besides the 4–velocity vector w_i, there are three (non–vectorial) measures of velocity relative to Q: the *3–velocity* v_α, more commonly referred to simply as the *velocity*, is defined to be

$$(v_1, v_2, v_3) := (w_1/w_0, w_2/w_0, w_3/w_0) .$$

A path through events $s_i \in S$ and $t_i \in T$ undergoes a displacement of d_{TS} in position space over the coordinate time interval $|t_0 - s_0|$: its *relative speed* is defined to be

$$|v| := d_{TS}/|t_0 - s_0| = \left(v_1^2 + v_2^2 + v_3^2\right)^{1/2}$$

and so Theorem 82(iii) implies that the relative speed is equal to the magnitude of the rectilinear directed velocity (defined in Section 7.5). Finally there is the *relative rapidity*

$$|r| := \tanh^{-1} |v| .$$

Relative speeds for paths have magnitude less than 1, while the speed of any lightlike line is equal to 1. Relative rapidities for paths are unbounded. The concept of rapidity is due to Robb (1911): for the case of one-dimensional motion, as we have seen in Chapter 8, directed rapidities are a "natural measure of speed" in the sense that they are arithmetically additive. The significance and properties of rapidity will be discussed in further detail after the proof of Theorem 87.

10. Standard model of Minkowski space–time

Now that we have established the isomorphism theorem we shall consider one fixed coordinate system and discuss the properties of space–time and the compositions of isotropy mappings with reference to this fixed coordinate system. Events will be specified by their coordinates and paths will be specified by the corresponding timelike lines. By considering the compositions of isotropy mappings we will be able to generate firstly the orthochronous (homogeneous) Lorentz group and secondly the orthochronous (inhomogeneous) Poincaré group. In Section 10.5 we define the set of "coordinate frames" and show that the transformations between them belong to the orthochronous Poincaré group.

The chapter concludes with Section 10.6 in which it is shown that the system of axioms is consistent. The mutual independence of the axioms will be demonstrated in the following Chapter 11.

10.1 Induced isotropy mappings are Lorentz transformations

The coordinatization of space–time results in equivalent algebraic conditions for induced isotropy mappings. These conditions are obtained in the next theorem and are expressed in terms of the *squared interval*

$$\| \Delta x \|^2 := \Delta x_0^2 - \Delta x_1^2 - \Delta x_2^2 - \Delta x_3^2 \ .$$

Affinities which leave the squared interval invariant are called *Poincaré transformations*, while those which also leave the origin $O(0,0,0,0)$ invariant are called *Lorentz transformations*.

Since the squared interval is not positive-definite, we shall only define the corresponding (non-squared) interval $\| \Delta x \|$ for vectors whose squared intervals are non-negative. Theorem 82 and the Isomorphism Theorem (Th.86) imply that the squared interval is (respectively) positive, negative, or zero for timelike, spacelike, and lightlike vectors (respectively). Accordingly lightlike vectors and lines are also known as *null vectors* and *null lines*.

In the next theorem we consider induced isotropy mappings which leave the event $O(0,0,0,0)$ invariant. Compositions of these isotropy mappings can then be used to generate some special types of mappings, which will be considered after the proof. It is convenient to define a matrix η_{ij} whose non–zero components are $\eta_{00} = 1$, $\eta_{11} = \eta_{22} = \eta_{33} = -1$.

Theorem 87 (Isotropy Mappings)

(i) *Induced isotropy mappings which leave the event $O(0,0,0,0)$ invariant are Lorentz transformations.*

(ii) *An affinity $x_i \rightarrow x'_i = a_{ij}x_j$ is a Lorentz transformation if and only if:*

$$a_{0i}a_{0j} - a_{1i}a_{1j} - a_{2i}a_{2j} - a_{3i}a_{3j} = \eta_{ij} .$$

(iii) *Induced isotropy mappings satisfy the additional condition:*

$$a_{00} \geq 1 .$$

(iv) *An affinity which has a fixed timelike line and satisfies the equations of (ii) and (iii) is an induced isotropy mapping.*

Remarks 1. Each Lorentz transformation $x_i \rightarrow x'_i = a_{ij}x_j$ has an inverse $x_i \rightarrow \bar{a}_{ij}x_j$ where $\bar{a}_{ij} = \eta_{il}a_{kl}\eta_{kj}$.

2. The set of equations corresponding to (ii) for the inverse transformation is

(ii′) $$a_{i0}a_{j0} - a_{i1}a_{j1} - a_{i2}a_{j2} - a_{i3}a_{j3} = \eta_{ij} .$$

Proof (i) An isotropy mapping θ is associated with an induced isotropy mapping θ^* which has an invariant timelike line and the induced affinity

$$x_i \rightarrow x'_i = a_{ij}x_j \tag{1}$$

sends a lightlike line

$$x_0^2 - x_1^2 - x_2^2 - x_3^2 = 0 \tag{2}$$

onto a lightlike line

$$(x'_0)^2 - (x'_1)^2 - (x'_2)^2 - (x'_3)^2 = 0 \tag{3}$$

Now if equations (1) are substituted into the quadratic form of (3) and terms are grouped we obtain

$$(x'_0)^2 - (x'_1)^2 - (x'_2)^2 - (x'_3)^2 =$$
$$(a_{0k}^2 - a_{1k}^2 - a_{2k}^2 - a_{3k}^2)x_k^2 + 2(a_{0k}a_{0l} - a_{1k}a_{1l} - a_{2k}a_{2l} - a_{3k}a_{3l})x_kx_l \quad (k < l). \tag{4}$$

We next show that

$$a_{0k}a_{0l} - a_{1k}a_{1l} - a_{2k}a_{2l} - a_{3k}a_{3l} = 0 \qquad (k < l) . \tag{5a}$$

First consider the two special cases of (2) with $(x_0, x_1, x_2, x_3) = (1, \pm 1, 0, 0)$. The resulting equations (3) and (4) then imply (5a) for the case $(k, l) = (0, 1)$ and the cases $(0, 2)$ and $(0, 3)$ are obtained similarly. Next the two special cases $(2, 1, \pm 1, 0)$ imply (5a) for the case $(k, l) = (1, 2)$ and the cases $(1, 3)$ and $(2, 3)$ are obtained similarly. Hence for a lightlike line (2), equations (3) and (4) become

$$(a_{0k}^2 - a_{1k}^2 - a_{2k}^2 - a_{3k}^2)x_k^2 = 0 .$$

If we consider the special cases of (2) with vectors $(1, 1, 0, 0)$, $(1, 0, 1, 0)$ and $(1, 0, 0, 1)$ we obtain (for $\alpha = 1, 2, 3$)

$$(a_{00}^2 - a_{10}^2 - a_{20}^2 - a_{30}^2) = -(a_{0\alpha}^2 - a_{1\alpha}^2 - a_{2\alpha}^2 - a_{3\alpha}^2) =: \lambda . \tag{5b}$$

Thus equations (4) and (5) imply that for any vector (x_0, x_1, x_2, x_3),

$$\begin{aligned}
(x_0')^2 - (x_1')^2 - (x_2')^2 - (x_3')^2 &= (a_{0k}a_{0l} - a_{1k}a_{1l} - a_{2k}a_{2l} - a_{3k}a_{3l})x_k x_l \\
&= \lambda(x_0^2 - x_1^2 - x_2^2 - x_3^2)
\end{aligned} \tag{5c}$$

For an isotropy mapping events on the invariant path are invariant, so for these events the two vectors are equal, whence

$$\lambda = 1 \tag{5d}$$

so the previous equation implies that the induced isotropy mapping θ^* is a Lorentz transformation which establishes (i).

(ii) Sufficiency is established by substituting the equations of (ii) into (4). To prove necessity we observe that, since Lorentz transformations leave the form $\| x \|^2$ invariant, any Lorentz transformation maps a lightlike line (2) onto a lightlike line (3). A similar argument shows that equations (5a,b,c) apply, with $\lambda = 1$ in equation (5c) by the definition of a Lorentz transformation. Equations (5a,b) with $\lambda = 1$ imply the equations of (ii).

A Lorentz transformation $x_i \to x_i' = a_{ij}x_j$ has an inverse with coefficients $\bar{a}_{ij} = \eta_{il}a_{kl}\eta_{kj}$ as is easily verified using equations (ii). Thus we have a set of equations which correspond to those of (ii) but with all subscripts transposed, which proves the two statements made as remarks. In particular we have

$$a_{00}^2 - a_{01}^2 - a_{02}^2 - a_{03}^2 = 1 . \tag{6}$$

181

(iii) Consider an invariant event (x_0, x_1, x_2, x_3) with $x_0 > 0$ on the fixed timelike line. For this invariant event $x_0 = a_{0j}x_j$ so that $x_0(1 - a_{00}) = a_{0\alpha}x_\alpha$. The Cauchy–Schwarz inequality implies that

$$x_0(1 - a_{00}) \le \left(a_{01}^2 + a_{02}^2 + a_{03}^2\right)^{1/2} \left(x_1^2 + x_2^2 + x_3^2\right)^{1/2} \quad .$$

Then equation (6) and the inequality for timelike lines imply that

$$x_0(1 - a_{00}) < \left(a_{00}^2 - 1\right)^{1/2} \left(x_0^2\right)^{1/2}$$

whence $(1 - a_{00})^2 < a_{00}^2 - 1$ which implies (iii).

(iv) The equations of (ii) imply (as in the derivation of (5c)) that $\| \, . \, \|^2$ is an invariant, so $\mathcal{V}$ is mapped bijectively. Now for an event with $x_0 > 0$,

$$x_0 \to x_0' = a_{00}x_0 + a_{01}x_1 + a_{02}x_2 + a_{03}x_3$$
$$\ge a_{00}x_0 - \left(a_{01}^2 + a_{02}^2 + a_{03}^2\right)^{1/2} \left(x_1^2 + x_2^2 + x_3^2\right)^{1/2}$$

by the Cauchy–Schwarz inequality and so equation (6) together with the inequality for timelike lines implies that $x_0' \ge a_{00}x_0 - (a_{00}^2 - 1)^{1/2} x_0$. Thus x_0 and its image x_0' are both positive and since $\| \, . \, \|^2$ is invariant, each event on the fixed timelike line is invariant. $\hfill q.e.d.$

This theorem implies the existence of two special types of isotropy mappings and two compositions of isotropy mappings which are called "Lorentz boosts".

A. Isotropy mappings with invariant path Q   The path Q has a 4–velocity vector $(1, 0, 0, 0)$: equations (ii) and (iii) of the preceding Theorem 87 imply that the transformation matrix has the form

$$a_{ij} = \begin{bmatrix} 1 & 0 \\ 0 & a_{\alpha\beta} \end{bmatrix}$$

where the sub-matrix $a_{\alpha\beta}$ represents an orthogonal transformation in the $X_1 X_2 X_3$–subspace. In this subspace the orthogonal transformations preserve the invariant quadratic form $x_1^2 + x_2^2 + x_3^2$ which is associated with a Euclidean metric $\sqrt{x_1^2 + x_2^2 + x_3^2}$ in position space. Isotropy mappings of this type will be called *orthogonal transformations*.

182

B. Isotropy mapping about the invariant path with 4–velocity $(\cosh r, \sinh r, 0, 0)$

A second special type of isotropy mapping has the transformation matrix

$$
a_{ij} = \begin{bmatrix}
\cosh 2r & -\sinh 2r & 0 & 0 \\
\sinh 2r & -\cosh 2r & 0 & 0 \\
0 & 0 & 1 & 0 \\
0 & 0 & 0 & 1
\end{bmatrix}
$$

which has an invariant path with 4–velocity vector $(\cosh r, \sinh r, 0, 0)$, as is readily verified by showing that equations (ii) and (iii) of Theorem 87 are satisfied and that the stated 4–velocity vector is invariant. This invariant path has a rapidity r in the X_1-direction which is equivalent to a speed $v = \tanh r$: the invariant 4–velocity vector may also be written as

$$
\left((1 - v^2)^{-1/2}, v(1 - v^2)^{-1/2}, 0, 0 \right) .
$$

C. Lorentz boost in X_1–direction with speed v

An important transformation which is not an isotropy mapping is the *Lorentz boost*. Consider the composition of two isotropy mappings of the type specified in the preceding paragraph but with rapidities in the X_1–direction of (firstly) $r/2$ and (secondly) r: the resulting transformation matrix is

$$
\lambda_{ij} = \begin{bmatrix}
\cosh 2r & -\sinh 2r & 0 & 0 \\
\sinh 2r & -\cosh 2r & 0 & 0 \\
0 & 0 & 1 & 0 \\
0 & 0 & 0 & 1
\end{bmatrix}
\begin{bmatrix}
\cosh r & -\sinh r & 0 & 0 \\
\sinh r & -\cosh r & 0 & 0 \\
0 & 0 & 1 & 0 \\
0 & 0 & 0 & 1
\end{bmatrix}
$$

$$
= \begin{bmatrix}
\cosh r & \sinh r & 0 & 0 \\
\sinh r & \cosh r & 0 & 0 \\
0 & 0 & 1 & 0 \\
0 & 0 & 0 & 1
\end{bmatrix} .
$$

This transformation matrix has no invariant timelike lines (the eigenvectors do not belong to $\mathcal{V}$). It maps each 4–velocity vector $(\cosh r', \sinh r', 0, 0)$ onto a 4–velocity vector $(\cosh[r + r'], \sinh[r + r'], 0, 0)$ and is called a "*Lorentz boost in the X_1-direction with rapidity r*" or a "*Lorentz boost in the X_1-direction with speed v*" where $v = \tanh r$. The Lorentz boost may also be written in terms of speed v as

$$
\lambda_{ij} = \begin{bmatrix}
(1 - v^2)^{-1/2} & v(1 - v^2)^{-1/2} & 0 & 0 \\
v(1 - v^2)^{-1/2} & (1 - v^2)^{-1/2} & 0 & 0 \\
0 & 0 & 1 & 0 \\
0 & 0 & 0 & 1
\end{bmatrix} .
$$

D. Boost in a general direction   The boost in the X_1–direction may be considered as a boost in the $(1, 0, 0)$ direction of position space. For a unit vector (d_1, d_2, d_3), with respect to the Euclidean metric of position space, we may define an orthogonal matrix $[g_{\alpha\beta}]$ such that $(g_{11}, g_{21}, g_{31}) = (d_1, d_2, d_3)$. Then the matrix $g \circ \lambda \circ g^{-1}$ represents a *boost of rapidity r and velocity v in the direction (d_1, d_2, d_3) of position space*.

E. Hyperbolic velocity space   The set of timelike lines $\mathcal{V}$ through the origin of $\mathbb{R}^4$ may be interpreted as the "points" of a geometry, which turns out to be a non–Euclidean hyperbolic or Bolyai–Lobachevskian geometry with relative rapidity as an intrinsic metric and a curvature of -1. The concept of rapidity was introduced by Robb (1911) who observed that directed rapidities gave an arithmetically additive, or "natural", measure of speed for the special case of rectilinear motion. This is described in detail in Chapter 8, especially Sections 8.1 and 8.4. Since the geometry is not flat, rapidities and 3–velocities do not combine as vectors. The geometry is known as "velocity space" or "rapidity space" in the literature and has been used by Boyer (1965), Fock (1964) and Smorodinsky (1965) to explain the relativistic phenomena of stellar aberration and Thomas procession. Isotropy mappings are isometries of rapidity space: the isotropy mapping of Section B (above) is a "reflection in a plane" while the Lorentz boosts, which were generated by the composition of two of these "reflection" mappings, are "translations" of rapidity space. These geometric interpretations are discussed in more detail by Robb (1911), Smorodinsky (1965), Yaglom (1969), Schutz (1973) and Yamasaki (1983).

10.2 The indefinite metric

In the next theorem we show that the *inner product*

$$x.y := x_0 y_0 - x_1 y_1 - x_2 y_2 - x_3 y_3$$

is invariant with respect to Lorentz transformations. The invariant squared interval is a special case of this inner product. Since the squared interval is not positive-definite, there are inequalities analogous to the usual Cauchy–Schwarz and Triangle Inequalities but with reversed sign for timelike and lightlike vectors.

Theorem 88 (Indefinite Metric)

(i) *Isotropy mappings which leave the origin $O(0,0,0,0)$ invariant leave both the inner product*

$$x.y = x_0 y_0 - x_1 y_1 - x_2 y_2 - x_3 y_3$$

and the squared interval

$$\| x \|^2 = x.x = x_0^2 - x_1^2 - x_2^2 - x_3^2$$

invariant. The invariant squared interval specifies an indefinite metric.

(ii) *"Reversed Cauchy–Schwarz Inequality" (indefinite metric).*
For two vectors which belong to distinct lines of $\overline{V}$,

$$| x.y | > \| x \| \, \| y \| \ .$$

Since M has an indefinite metric, the inequality is reversed as compared with the usual Cauchy–Schwarz inequality for positive definite metric spaces.

(iii) *"Reversed Triangle Inequality" (indefinite metric).*
For two vectors x, y which belong to distinct lines of $\overline{V}$,

$$\| x + y \| > \| x \| + \| y \| \ .$$

This inequality is also reversed as compared with the usual triangle inequality for positive definite metric spaces.

Proof Consider the inner product form (for arbitrary vectors x, y)

$$x.y = (x_0, x_1, x_2, x_3).(y_0, y_1, y_2, y_3) = x_0 y_0 - x_1 y_1 - x_2 y_2 - x_3 y_3$$

and the expansion

$$(x - y).(x - y) = x.x - 2x.y + y.y \ .$$

It follows that

$$2x.y = \| x \|^2 + \| y \|^2 - \| x - y \|^2 \ .$$

Since all terms on the right hand side are invariant quadratic forms, the inner product on the left hand side is invariant.

(ii) To establish the "Reversed Cauchy–Schwarz inequality" we observe that for two non-parallel timelike or lightlike vectors there are two linear combinations which are parallel to lightlike lines of the boundary $\partial\mathcal{V}$, so there are two distinct solutions (for μ) of the equation

$$(\mu x + y).(\mu x + y) = \mu^2 x.x + 2\mu x.y + y.y = 0$$

and so the discriminant $4[(x.y)^2 - (x.x)(y.y)]$ is positive, which implies the inequality.

(iii) Next we use the Reversed Cauchy–Schwarz inequality to obtain the Reversed Triangle Inequality. Now

$$\begin{aligned}
\parallel x + y \parallel^2 &= (x+y).(x+y) \\
&= \parallel x \parallel^2 + 2x.y + \parallel y \parallel^2 \\
&> \parallel x \parallel^2 + 2 \parallel x \parallel \parallel y \parallel + \parallel y \parallel^2
\end{aligned}$$

since x and y are not parallel, hence

$$\parallel x + y \parallel^2 > (\parallel x \parallel + \parallel y \parallel)^2$$

from which the Reversed Triangle Inequality (part(iii)) follows. $\qquad\qquad q.e.d.$

This theorem illustrates some of the important differences between a space–time with an indefinite metric (or pseudo–metric) and a geometry (which has a positive–definite metric).

Optical lines The "Reversed Triangle Inequality" (part(iii)) implies that optical lines (see Section 6.4) correspond to limiting lines: that is, an optical line is a straight line whose speed is unity. The question raised in Section 6.4 can now be answered: the only sets of events which are in optical line correpond to subsets of limiting lines.

10.3 The orthochronous Lorentz group

A Poincaré transformation

$$x_i \longrightarrow x_i' = a_{ij}x_j + b_j$$

is said to be *orthochronous* if a_{00} is positive. (It will be shown in Theorem 90 that orthochronous Poincaré transformations preserve "causal relations".) Theorem 87 implies that isotropy mappings with invariant event $O(0,0,0,0)$ are orthochronous Lorentz transformations, as are the orthogonal transformations and Lorentz boosts. In the next theorem we will show that the orthochronous Lorentz transformations form a sub-group known as the *orthochronous Lorentz group*.

Theorem 89 (Orthochronous Lorentz Group)

(i) *The set of orthochronous Lorentz transformations is a group.*

(ii) *Any orthochronous Lorentz transformation*

$$x_i \longrightarrow x_i' = a_{ij} x_j$$

has an inverse transformation $x_i \longrightarrow \bar{a}_{ij} x_j$ *with coefficients* $\bar{a}_{ij} = \eta_{il} a_{kl} \eta_{kj}$.

(iii) *Any orthochronous Lorentz transformation may be represented as a composition of isotropy mappings, or as a composition of a Lorentz boost followed by an orthogonal transformation. Conversely, any composition of isotropy mappings (which leave the origin $O(0,0,0,0)$ invariant) is an orthochronous Lorentz transformation.*

(iv) *For any given event $e(e_0, e_1, e_2, e_3)$, the set of orthochronous Lorentz transformations which leave e invariant is isomorphic to the orthochronous Lorentz group. For each orthochronous Lorentz transformation a, there is a corresponding transformation $\tau^{-1} \circ a \circ \tau$ which leaves e invariant, where τ is the translation*

$$\tau : \ x_i \longrightarrow x_i' = x_i - e_i \ .$$

Remarks   1. The orthochronous Lorentz group includes the special transformations of Section 10.1(A–D).

2. Part (iv) applies in particular to induced isotropy mappings.

Proof (ii) This is the "Remark" attached to Theorem 87 (and has been proved in the paragraph which precedes the proof of part (iii) of that theorem).

(i) By definition, the composition of Lorentz transformations is a Lorentz transformation and the properties of matrix multiplication imply the associative property. By part (ii) each orthochronous Lorentz transformation has an inverse which is also an orthochronous Lorentz transformation. Finally the composition of orthochronous Lorentz transformations

$$x_i \longrightarrow x_i' = a_{ij} x_j \quad \text{and} \quad x_i' \longrightarrow x_i'' = b_{ij} x_j'$$

is

$$x_i \longrightarrow x_i'' = c_{ij} x_j \quad \text{where} \quad c_{ij} = b_{ik} a_{kj}$$

so that

$$c_{00} \geq (b_{00}, |\, b_{01}\,|, |\, b_{02}\,|, |\, b_{03}\,|) \cdot (a_{00}, |\, a_{10}\,|, |\, a_{20}\,|, |\, a_{30}\,|) \,,$$

since by Theorem 87((ii),(ii′)) both vectors are timelike. Now the "Reversed Cauchy–Schwarz" inequality (of the previous theorem) and equations ((ii),(ii′)) of Theorem 87

imply that
$$c_{00} \geq 1^{1/2} \cdot 1^{1/2}$$

and so $c_{00} \geq 1$ and the composition of transformations is orthochronous.

(iii) Let a_{ij} be the transformation matrix of an orthochronous Lorentz transformation. Let $r := \cosh^{-1} a_{00}$: then the equations of Theorem 87(ii) and equations (1) imply that

$$a_{10}^2 + a_{20}^2 + a_{30}^2 = \sinh^2 r \ .$$

Now take an orthogonal matrix $\rho_{\alpha\beta}$ such that

$$\rho_{\alpha\beta} : \ (1,0,0) \longrightarrow \frac{1}{\sinh r}(a_{10}, a_{20}, a_{30})$$

and define the orthogonal transformation ρ by the matrix

$$\rho_{ij} := \begin{bmatrix} 1 & 0 \\ 0 & \rho_{\alpha\beta} \end{bmatrix} \ .$$

Let λ_{ij} be the transformation matrix of a Lorentz boost λ of rapidity r in the X_1–direction. Then

$$\lambda : \ (1,0,0,0) \longrightarrow (\cosh r, \sinh r, 0, 0) = \left(a_{00}, \left(a_{10}^2 + a_{20}^2 + a_{30}^2\right)^{1/2}, 0, 0\right)$$

whence
$$\rho \circ \lambda : \ (1,0,0,0) \longrightarrow (a_{00}, a_{10}, a_{20}, a_{30}) \ .$$

But also
$$a : \ (1,0,0,0) \longrightarrow (a_{00}, a_{10}, a_{20}, a_{30})$$

so the transformation $a^{-1} \circ \rho \circ \lambda$ leaves $(1,0,0,0)$ invariant and is therefore an orthogonal transformation μ, whence

$$a = \rho \circ \lambda \circ \mu^{-1}$$
$$= (\rho \circ \mu^{-1}) \circ (\mu \circ \lambda \circ \mu^{-1})$$

where $(\mu \circ \lambda \circ \mu^{-1})$ is a Lorentz boost and $(\rho \circ \mu^{-1})$ is an orthogonal transformation. This completes the proof of the first proposition of (iii). The converse proposition is a consequence of (i) and the previous Theorem 87(i).

(iv) This is an immediate consequence of the affine structure of $\mathcal{M}$ and M which was established in the Isomorphism Theorem 86.   *q.e.d.*

10.4 The orthochronous Poincaré group of motions

The temporal order relation $<$ was first defined (in Section 5.2) for a compact collinear set and then (in Section 5.5) was extended so as to apply to any pair of events which both belong to the same collinear set. We will now extend the definition to apply to any pair of events and then we will show that isotropy mappings preserve the relation of temporal order and generate the orthochronous Poincaré group. The *extended temporal order relation* $<$ is defined :

$$x < y \iff \| y_i - x_i \|^2 > 0 \text{ and } x_0 < y_0 .$$

This relation is an extension of the previous temporal order relation. It is clearly irreflexive and asymmetric and it is transitive as a consequence of the "reversed triangle inequality" of Theorem 88(iii), so it is a partial order relation. Any automorphism of space–time which preserves the relation of temporal order is said to be *causal*.

So far we have concentrated our attention on those isotropy mappings which leave the event $O(0,0,0,0)$ invariant. Theorem 89(iv) enables us to consider all isotropy mappings, so we will now compose isotropy mappings to generate "spacelike reflections", "spacelike translations", "timelike translations" and more general "translations". That is, we will show that any translation may be generated by composing isotropy mappings: this result will be used in Theorem 90 to discuss the orthochronous Poincaré group.

A. Spacelike reflections   The orthogonal transformation (isotropy mapping)

$$\rho : \begin{bmatrix} x_0 \\ x_1 \\ x_2 \\ x_3 \end{bmatrix} \longrightarrow \begin{bmatrix} x'_0 \\ x'_1 \\ x'_2 \\ x'_3 \end{bmatrix} = \begin{bmatrix} 1 & 0 & 0 & 0 \\ 0 & -1 & 0 & 0 \\ 0 & 0 & 1 & 0 \\ 0 & 0 & 0 & 1 \end{bmatrix} \begin{bmatrix} x_0 \\ x_1 \\ x_2 \\ x_3 \end{bmatrix}$$

is called a *spacelike reflection in the X_2X_3–plane* (that is, in the $x_1 = 0$ plane of the $X_1X_2X_3$–position subspace). The previous Theorem 89(iv) implies that there is an orthogonal transformation (isotropy mapping) called a *spacelike reflection in the $x_1 = d$ plane* which may be defined, using the translation

$$\tau : \begin{bmatrix} x_0 \\ x_1 \\ x_2 \\ x_3 \end{bmatrix} \longrightarrow \begin{bmatrix} x'_0 \\ x'_1 \\ x'_2 \\ x'_3 \end{bmatrix} = \begin{bmatrix} x_0 \\ x_1 \\ x_2 \\ x_3 \end{bmatrix} - \begin{bmatrix} 0 \\ d/2 \\ 0 \\ 0 \end{bmatrix}$$

to be

$$\rho_d := \tau^{-1} \circ \rho \circ \tau \ .$$

Then

$$\rho_d : \begin{bmatrix} x_0 \\ x_1 \\ x_2 \\ x_3 \end{bmatrix} \longrightarrow \begin{bmatrix} x_0 \\ -x_1 + d \\ x_2 \\ x_3 \end{bmatrix}$$

and ρ_d is an isotropy mapping whose invariant path has position space coordinates $(d, 0, 0)$.

B. Space translation   The mapping

$$\delta_d := \rho_d \circ \rho$$

generated by the composition of two isotropy mappings is the affinity

$$\delta_d : \begin{bmatrix} x_0 \\ x_1 \\ x_2 \\ x_3 \end{bmatrix} \longrightarrow \begin{bmatrix} x_0 \\ x_1 \\ x_2 \\ x_3 \end{bmatrix} + \begin{bmatrix} 0 \\ d \\ 0 \\ 0 \end{bmatrix}$$

and is called a *space translation of d in the X_1–direction*.

C. Time translation   Given any real number t we can define a Lorentz boost λ of rapidity $r = \sinh^{-1} t$ in the X_1–direction and then (with $c := \cosh r$), the mapping

$$\tau_t := \lambda^{-1} \circ \delta_{-1} \circ \lambda \circ \delta_c$$

which is composed entirely from isotropy mappings (six of them) has the effect

$$\tau_t : \begin{bmatrix} x_0 \\ x_1 \\ x_2 \\ x_3 \end{bmatrix} \longrightarrow \begin{bmatrix} x_0 \\ x_1 \\ x_2 \\ x_3 \end{bmatrix} + \begin{bmatrix} t \\ 0 \\ 0 \\ 0 \end{bmatrix}$$

(as may be easily verified) and is called a *time translation mapping*. It is also possible to specify τ_t using only four isotropy mappings.

D. General translation mapping   By composing space translation mappings with orthogonal transformations it is possible to generate space translation mappings for any specified space displacement.   It is then possible to compose with time translation mappings to generate translation mappings for any specified displacement.   Thus *any translation may be generated by the composition of* *isotropy mappings*. Translations are known, respectively, as "timelike", "spacelike" and "lightlike".

E. The orthochronous Poincaré group In the next theorem we show firstly that isotropy mappings generate the (full inhomogeneous) orthochronous Poincaré group of motions, which is the largest causal subgroup of the Poincaré group, and secondly that Minkowski space–time is isotropic.

Theorem 90 (Orthochronous Poincaré Group)

(i) *An orthochronous Poincaré transformation may be represented as the composition of an orthochronous Lorentz transformation and a translation mapping.*

(ii) *Orthochronous Poincaré transformations are causal and form a group known as the "orthochronous Poincaré group". Conversely any causal Poincaré transformation is orthochronous.*

(iii) *An orthochronous Poincaré transformation may be represented as a composition of isotropy mappings, or as a composition of Lorentz boosts and orthogonal transformations. Conversely, any composition of isotropy mappings belongs to the orthochronous Poincaré group.*

Proof (i) Any affinity $x_i \longrightarrow x_i'$ may be represented as the composition of a homogeneous transformation and a translation; that is

$$x_i' = a_{ij}x_j + b_i \ .$$

(ii) Thus for $x_i < y_i$,

$$y_0' - x_0' = a_{0j}(y_j - x_j)$$
$$= a_{00}(y_0 - x_0) + a_{01}(y_1 - x_1) + a_{02}(y_2 - x_2) + a_{03}(y_3 - x_3)$$

and, by the usual Cauchy–Schwarz inequality,

$$y_0' - x_0' \geq a_{00}(y_0 - x_0)-$$
$$- \left(a_{01}^2 + a_{02}^2 + a_{03}^2\right)^{1/2} \left((y_1 - x_1)^2 + (y_2 - x_2)^2 + (y_3 - x_3)^2\right)^{1/2}$$
$$\geq a_{00}(y_0 - x_0) - (a_{00}^2 - 1)^{1/2}(y_0 - x_0)$$

using the equations of Theorem 87(ii) and the definition of temporal order, so

$$y_0' - x_0' \geq (y_0 - x_0)\left(a_{00} - (a_{00}^2 - 1)^{1/2}\right)$$

from which we see that both $y_0' - x_0' > 0$ and $y_0 - x_0 > 0$ if and only if $a_{00} \geq 1$. That is, Poincaré transformations are causal if and only if they are orthochronous. The group property is an immediate consequence of (i) above and Theorem 89(i).

(iii) Both the orthochronous homogeneous transformation and the translation of (i) can be composed from isotropy mappings, or from Lorentz boosts and orthogonal transformations as shown in the preceding Theorem 89 and in the discussion of translation mappings which preceded this theorem. To establish the converse proposition, we observe that any isotropy mapping may be expressed in the form $\delta^{-1} \circ a \circ \delta$, where δ is a translation and a is an isotropy mapping with fixed event $O(0,0,0,0)$: since δ^{-1}, a, and δ are orthochronous, part (ii) above implies that the given isotropy mapping is orthochronous. Again by part (ii) above, it follows that any composition of isotropy mappings is orthochronous.   *q.e.d.*

This theorem completes our discussion of Minkowski space–time and the automorphisms generated by isotropy mappings.

10.5 Transformations between coordinate frames

At this stage it is worth remarking that only one coordinate system has been specified. Induced isotropy mappings may also be considered to map this space–time coordinate system onto new space–time coordinate systems. Accordingly, the previous theorem implies the existence of other coordinate systems and this result may be stated as:

Theorem 91 (Existence of Coordinate Frames)
For any timelike line and any event on the line, there is a space–time coordinate system which has the line and the event, respectively, as the origins of position–space and space–time.

Remarks   1. A coordinate system obtained in this way is called a *coordinate frame*.
2. The position–space is Euclidean and the previous concepts of 4–velocity, 3–velocity and rapidity apply, so the speed of lightlike lines (usually called the "speed of light" in the literature) is constant and equal to unity in every coordinate frame[1].
3. In fact for any given timelike line and any event on the line, there is a set of coordinate frames related by orthogonal transformations (as described in Section 10.1).

Theorem 90 now implies:

Theorem 92 (Transformations Between Coordinate Frames)
Any two coordinate frames are related by an orthochronous Poincaré transformation.

The procedure of coordinatization involved the selection of any path with any affine parametrisation. Since an affine or natural time scale is only defined up to an arbitrary change of algebraic sign and an arbitrary change of scale (as well as an arbitrary change of origin), there can be many sets of cooordinate frames. Any two sets of coordinate frames are related by a change of scale either with or without a change of algebraic sign. Thus the preceding theorem and the Isomorphism Theorem 86 imply:

Theorem 93 (Representation Theorem)

(i) *Any model of Minkowski space–time $\mathcal{M}$ may be represented by the usual coordinate model.*

(ii) *The correspondence between any two coordinate representations is specified by a Poincaré transformation (which could be non-orthochronous) and a change of scale.*

10.6 Consistency of the system of axioms

To show that the system of axioms is consistent, it must be shown that each of the axioms is satisfied by the model M (which is isomorphic to $\mathcal{M}$). The model M is defined on $\mathbb{R}^4$ and hence its consistency ultimately depends on the consistency of the reals. We can show that each of the axioms is satisfied by the model M: this is left as an exercise for the reader[2]. Thus we can demonstrate the consistency of the axioms relative to the consistency of the reals.

11. Independence models

In this chapter we demonstrate the independence of the axioms. For each axiom, say Axiom X, we describe a *model* or *interpretation*

$$M_X = \langle \mathcal{E}_X, \mathcal{P}_X, [\ldots] \rangle$$

in which all axioms, other than the given Axiom X, are satisfied. The existence of this model M_X demonstrates that Axiom X can not be deduced as a theorem from the remaining axioms. It is in this sense that the given Axiom X is independent of all the other axioms, so the model is called an *independence model*. In order to demonstrate that we have an *independent axiomatic system*, we will describe an independence model for each axiom of the system.

Since there are fifteen axioms altogether, we will describe fifteen independence models, one for each axiom, together with additional independence models for the first and fourth Axioms of Order (Axioms O1 and O4), the fourth Axiom of Incidence (Axiom I4) and the Axiom of Continuity (Axiom C). These additional independence models serve, by their contrasts with Minkowski space–time, to illustrate the rôles played by the corresponding (non–negated) axioms in determining the structure of Minkowski space–time.

Hilbert (1899, 1913) described independence models which established the independence of some, but not all, of the axioms of his system. Some of our independence models resemble models used by Hilbert. An independent and categorical axiom system for affine geometry was given by Veblen (1904) who showed that Euclidean geometry[1] may be defined on an affine space, with reference to an elliptic polar system in the improper plane "at infinity" and a class of perpendicular reflections in ordinary planes. Moore[2] (1908) extended this work of Veblen by appending five axioms of congruence and developed two independent and categorical systems of axioms for Euclidean geometry, each system containing sixteen axioms.

Veblen and Young (1908, 1918) have described an independent and categorical axiomatic system for real projective geometry using eleven axioms.

11.1 Independence models M_{O1} and M'_{O1}

The first model M_{O1} is obtained by taking the usual cartesian model for Minkowski space–time and defining the obvious relation of betweenness for triples of events on each straight line; that is, on lightlike and spacelike lines as well as on the timelike lines. The second model M'_{O1} is also obtained by taking the usual cartesian model for Minkowski space–time but for this model the relation of betweenness is defined for triples of events which need not be collinear:

$$[xyz] \iff x < y < z \quad \text{or} \quad x > y > z \,,$$

where $<$ denotes the extended temporal order relation (of Section 10.4) This relation of betweeenness may be considered as a form of "causal" betweenness.

Both models fail to satisfy Axiom O1, but do satisfy all the remaining axioms.

11.2 Independence model M_{O2}

Take the standard cartesian model for Minkowski space–time and, for each timelike line and for each triple of events a, b, c on the timelike line, specify the betweenness relation in terms of the time coordinate x_0 such that

$$[abc] \iff a_0 < b_0 < c_0 \quad .$$

Then this model does not satisfy Axiom O2 but it does satisfy all the other axioms.

11.3 Independence model M_{O3}

Take the standard cartesian model for Minkowski space–time and for each timelike line and for each triple of events a, b, c on the timelike line, we specify that

$$[abc] \iff a_0 < b_0 < c_0 \quad \text{or} \quad a_0 > b_0 > c_0 \quad \text{or} \quad a = b = c \,.$$

11.4 Independence models M_{O4} and M'_{O4}

The sets of events and paths for model M_{O4}   The independence model M_{O4} is a subset of the four–dimensional projective space P^4. First consider the non–degenerate ruled quadric

$$x_1^2 + x_2^2 + x_3^2 - x_4^2 - x_5^2 = 0 \qquad (\mathcal{Q})$$

which is illustrated for non–zero values of x_5 in Figure 89 by displaying the hyperboloid (with the representative non–zero value of $x_5 = 1$)

$$x_1^2 + x_2^2 + x_3^2 - x_4^2 = 1 \qquad (\mathcal{H})$$

where the X_3–axis is implied in the illustration (and points "at infinity" lie on the hyperplane at infinity where $x_5 = 0$).

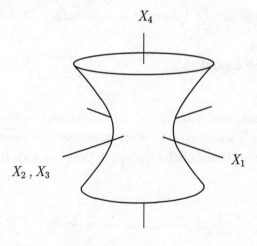

X_4

X_2 , X_3

X_1

Figure 89

In this illustration, vertical lines meet at the point at infinity in the X_4–direction; that is, at $(0, 0, 0, 1, 0)$ in P^4. Thus tangents from this point meet the hyperboloid $\mathcal{H}$ and the quadric $\mathcal{Q}$ at points of a non–degenerate non–ruled quadric

$$x_1^2 + x_2^2 + x_3^2 - x_5^2 = 0 \,,$$

which is illustrated in Figure 89 as the intersection of $\mathcal{H}$ with the "base plane". The lines which do not intersect the quadric $\mathcal{Q}$ satisfy the inequality

$$x_1^2 + x_2^2 + x_3^2 - x_5^2 < 0 \,.$$

Horizontal lines parallel to the X_1–axis meet at the point at infinity in the X_1–direction; that is, at $(1,0,0,0,0)$ in P^4. Thus tangents from this point meet the quadric Q at points of the non–degenerate ruled quadric

$$x_2^2 + x_3^2 - x_4^2 - x_5^2 = 0$$

and the lines which do not intersect the quadric Q satisfy the inequality

$$x_2^2 + x_3^2 - x_4^2 - x_5^2 > 0 .$$

We have now shown that there are two types of points of P^4 — the former whose tangents meet the quadric Q in a non–degenerate non–ruled quadric will not be part of our model, while the latter whose tangents meet the quadric Q in a non–degenerate ruled quadric will be part of the model. If the illustration is thought of as a hyperboloid of revolution then the excluded points are all those on its surface and in its "throat". The paths are lines of P^4 which do not meet the excluded region, so each path has the topology of a circle. The hyperboloid $\mathcal{H}$ illustrated in Figure 89 is the intersection of the quadric Q with $\mathbb{R}^4$ but in P^4 the surface is the quadric Q and the illustration of Figure 89 corresponds to $x_5 = 1$. Thus the quadric Q has the topology $S^2 \times S^1$ and the space of paths through any fixed event is homeomorphic to the interior of the double covering space of $S^1 \times S^1$, (that is, the interior of the double covering space of a torus) as may be easily verified by considering the special case of the fixed event $(x_1, x_2, x_3, x_4, x_5) = (1,0,0,0,0)$. Thus this independence model is not even "locally Minkowskian" . (A second independence model M'_{O4} which is "locally Minkowskian" is based on the set of points within the "throat" and will be described after the discussion of the present model M_{O4}).

The betweenness relation   The relation of betweenness is defined on the set of events by first defining it for triples of events on each path. We observe that each path lies outside the hyperboloid $\mathcal{H}$ (Figure 89). Now consider the non–degenerate non–ruled quadric

$$y_1^2 + y_2^2 + y_3^2 + y_4^2 = y_5^2 \qquad (S^3)$$

which, for the affine sub–space $\mathbb{R}^4$ (of P^4 with $y_5 = 1$), is a 3–sphere with centre at the origin of $\mathbb{R}^4$. For each event of the model M_{O4} there is a unique line through the origin of $\mathbb{R}^4$ and this line maps the given event onto a pair of antipodal points on the 3–sphere S^3. Each path of the model is mapped onto a great circle on the 3–sphere S^3. If we now equip the affine space $\mathbb{R}^4$ (with $x_5 = 1$) with the usual

Euclidean metric, we can specify a measure of "distance" between any given pair of events by taking the smaller of the two angles between the corresponding lines from the origin to the events (or by taking a right angle where the angles are equal). The betweenness relation is defined to apply only to triples of distinct events:

$$[abc] \iff \begin{cases} a, b, c \text{ are distinct and} \\ \max\{d(a,b), d(b,c)\} \leq \max\{d(a,b), d(b,c), d(a,c)\} \end{cases}.$$

where $d(\cdot, \cdot)$ denotes the distance function. Thus any three events (points) on the same path can be labelled as a, b, c such that either: (i) $[abc]$, or (ii) $[abc]$ and $[bac]$, or (iii) $[abc]$ and $[bac]$ and $[acb]$.

The axioms of order   Axioms O1,O2,O3 and O5 are satisfied as is easily verified.

To see that Axiom O4 is not satisfied in this model, consider four events a, b, c, d on the same path such that the corresponding points on the corresponding great circle are separated by angles of $\pi/4$. Thus we have $[abc]$ and $[bcd]$ (and $[dab]$) but not $[abd]$ which demonstrates that Axiom O4 does not apply.

Now consider the configuration of the given conditions of Axiom O6: this configuration spans a two dimensional projective plane in which all paths are projective lines which do not meet the excluded set of points, so Axiom O6 is satisfied in this model since it is satisfied in the projective plane.

The axioms of incidence   It is easily verified that all axioms of incidence are satisfied in this model.

M_{O4} **satisfies the Axiom of Isotropy**   The subspace spanned by Q, R, S is either three–dimensional or two–dimensional. For the three–dimensional case this subspace meets the ruled quadric Q in a ruled quadric surface Q_2. The paths Q, R, S meet at a point x which is not on Q_2. The polarity corresponding to the quadric surface Q_2 relates the point x to a polar plane Π which meets Q_2 in a conic C. Since $x \notin Q_2$ the polar plane Π does not contain x so it meets the paths Q, R, S in points q, r, s (respectively) which are no–tangent points of the conic C (Figure 90). The lines qr and qs meet the conic C in points r_1, r_2 and s_1, s_2 such that $[r_1, q, r, r_2]$ and $[s_1, q, s, s_2]$, where $[\cdot \cdot \cdot \cdot]$ denotes the usual four–point order relation of projective geometry. Let $w := r_1 s_2 \cap s_1 r_2$ and let $z := r_1 s_1 \cap r_2 s_2$ and let P^2 be the polar plane of z with respect to the quadric surface Q_2.

Since the point z is in the polar plane Π of the event x (with respect to the quadric surface Q_2), the event x is in the polar plane P^2 of the point z (with respect

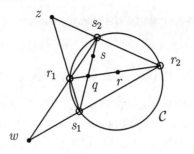

Figure 90

to the quadric surface $\mathcal{Q}_2$). Also the point q is on the polar line of the point z (with respect to the conic $\mathcal{C}$), so the point q is on the polar plane P^2 of the point z (with respect to the quadric surface $\mathcal{Q}_2$). Since both x and q are in P^2 the path $Q := xq$ is in P^2, the polar plane of z with respect to the quadric surface $\mathcal{Q}_2$, and therefore the path Q is in P^3, the polar hyperplane of the point z with respect to the quadric $\mathcal{Q}$. Now the harmonic homology with centre z and axial hyperplane P^3 maps the quadric $\mathcal{Q}$ bijectively onto itself[3], maps the events of the model M_{O4} bijectively onto itself, maps the set of paths bijectively onto itself, leaves the events of the path Q invariant, and interchanges the planes $pl[Q, R]$ and $pl[Q, S]$. Thus the harmonic homology satisfies the conditions stated in the Axiom of Isotropy (Axiom S).

In the case where the paths Q, R, S span a two–dimensional subspace take any three dimensional subspace which contains this two–dimensional subspace. Then the three dimensional subspace meets the quadric $\mathcal{Q}$ in a quadric surface $\mathcal{Q}_2$ and the polar plane Π and points q, r, s can be defined as before. Let the point z be defined as the harmonic conjugate of q with respect to the points of intersection of the line $qr \ (= qs)$ with the conic $\mathcal{C}$. Then, using the same argument as for the previous case, the point z has a polar hyperplane P^3 which contains the path Q and the harmonic homology with centre z and axial hyperplane P^3 satisfies the conditions stated in the Axiom of Isotropy.

M_{O4} **satisfies the Axiom of Continuity** For this model, the betweenness relation $[\cdots]$ is defined in terms of the distance function $d(\cdot, \cdot)$. If this distance function is replaced by an angle measure

$$\alpha(\cdot, \cdot) := 2d(\cdot, \cdot)$$

then each event, which has already been mapped onto a line through the origin, is now mapped onto a single point on a circle (rather than a pair of points) and the angle measure $\alpha(\cdot, \cdot)$ corresponds to the usual measure of angle from the centre.

Define the set of bounds of Q_1 as

$$\mathcal{B}_1 := \{Q_b : [Q_0 \; Q_1 \; Q_b]; \; Q_b \in Q\} \; .$$

Let Q_d be the point diametrically opposite Q_0. If $\mathcal{B}$ is non–empty then $\mathcal{B}_1$ is also non–empty and contains Q_d. Let Q_c be the bisector of the longer arc Q_0Q_1 and let Q_1' be the reflection of Q_1 in Q_0. If $\alpha(Q_0, Q_1) \leq 2\pi/3$, then $\mathcal{B}_1$ consists of all points on the shorter arc $(Q_1, Q_c]$. If $\alpha(Q_0, Q_1) > 2\pi/3$, then $\mathcal{B}_1$ consists of all points on the shorter arc $(Q_1, Q_1']$. Sets of bounds $\mathcal{B}_2, \mathcal{B}_3, \ldots$ are defined similarly and $\mathcal{B} \subseteq \cap_n \mathcal{B}_n$.

Let $\mu := \inf\{\alpha(Q_n, Q_d)\}$. If $\mu \geq \pi/3$ then all the Q_n are contained in the longer arc $[Q_0 Q_c)$ which also contains the closest bound of $\mathcal{B}$ at an angle of $\pi - \mu$ from Q_0. If $\mu < \pi/3$ then all the Q_n are further than μ from Q_d and $\mathcal{B}$ is a closed arc which extends on both sides of Q_d by the angle μ: both end–points of this arc are closest bounds.

Independence model M_{O4}'   This model is obtained in much the same way as the previous model, except that the set of events for model M_{O4}' is the set of points within the "throat" of the quadric $\mathcal{Q}$ and the paths are lines within the throat which do not meet the quadric. Each line has the topology of a circle and the model is temporally orientable.

A relation of betweenness may be defined on each path using the sphere $\mathcal{S}^3$ and a similar definition as for the previous model M_{O4}.

The space of paths through any fixed event has the topology of the interior of a 2–sphere, so the model M_{O4}' is locally Minkowskian. The model satisfies all axioms, apart from Axiom O4, as can be demonstrated in much the same way as for the previous model.

11.5 Independence model M_{O5}

This model is a complex Minkowski space–time over $\mathbb{C}^4$ with the *invariant squared interval* between events $w = (w_0, w_1, w_2, w_3)$ and $z = (z_0, z_1, z_2, z_3)$ defined by

$$\|w - z\|^2 := (w_0 - z_0)(w_0^* - z_0^*) - \sum_{i=1}^{3}(w_i - z_i)(w_i^* - z_i^*) \; .$$

(where the asterisk denotes complex conjugation). Given any event b and any four–vector $u \; (= (u_0, u_1, u_2, u_3))$ such that $\|u\|^2 > 0$, there is some *path*

$$Q = \{b + uz : z \in \mathbb{C}\}$$

through b in the direction u, where u is called the 4–velocity and the triple $(u_1/u_0, u_2/u_0, u_3/u_0)$ is called the 3–velocity or (more simply) just the velocity. Each path is isomorphic to the complex plane $\mathbb{C}$. The betweenness relation is defined so that $[abc]$ if and only if:

(i) a, b, c are distinct events which belong to one path, and

(ii) $\|a - b\| + \|b - c\| = \|a - c\|$.

In addition to the familiar transformations of the inhomogeneous Poincaré group, there are phase transformations — for any quadruple $(\theta_0, \theta_1, \theta_2, \theta_3)$ there is a phase transformation

$$z_k \mapsto z_k e^{i\theta_k} \qquad\qquad (k = 0, 1, 2, 3)$$

(where there is no sum over the index k).

With this enlarged group of interval–preserving transformations any pair of events of any given path can be mapped — one onto the origin and the other onto an event $(x, 0, 0, 0)$ where x is real. Thus given any three distinct paths which meet at some event, the common event can be mapped onto the origin, one path can be mapped onto the Z_0–axis in such a way that any other given event of the path can be mapped onto the positive part of the real sub–axis of the Z_0–axis, one of the other paths can be mapped into the $Z_0 Z_1$–subspace, and the remaining path can be mapped into the $Z_0 Z_1 Z_2$–subspace (or the $Z_0 Z_1$–subspace if the three paths are linearly dependent).

By considering the interval between the event $e(0, 1, 0, 0)$ and a typical event $(z, 0, 0, 0)$ on the path Q coincident with the Z^0(plane)–axis, we see that

$$Q(e, \emptyset) = \{(z, 0, 0, 0) : zz^* \leq 1\} ;$$

that is the unreachable set is a closed circular region. It is easy to convince oneself (by considering the group of all interval–preserving mappings — that is, isometries) that the closed circular shape applies to every unreachable set.

The axioms of order   Each path is isomorphic to the complex plane and three events (of a path) satisfy the definition of betweenness if they lie on a straight line (in the complex plane). Thus all axioms of order, except for Axiom O5, are satisfied.

The axioms of incidence   All axioms of incidence are satisfied: to verify the Axiom of Uniqueness (Axiom I3) it is sufficient to show that each path can be specified with a unique 3–velocity — the other axioms of incidence are easily verified for this model.

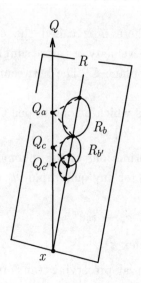

Figure 91

The Axiom of Isotropy   Next consider two paths Q, R as in the statement of the conditions for the Axiom of Isotropy (Axiom S). It is clear that the transformations previously described are sufficient to bring the event of coincidence x to the origin, the path Q to the real part of the Z_0–axis, the path R to the $Z_0 Z_1$–subspace with positive real 3–velocity $(v, 0, 0)$ and the event Q_a to the positive half–line of the real sub–axis of the Z_0–axis: so without loss of generality we consider this latter special case and we suppose for convenience that a is the Z_0–coordinate of the event Q_a (Figure 91). Let R_b be the closest event of $R(Q_a, \emptyset)$ to x and let Q_c be the closest event of $Q(R_b, \emptyset)$ to x (on the positive real sub–axis): then for any $Q_{c'}$ with $[x, Q_{c'}, Q_c]$ a consideration of similar circles and the lines tangent to them shows that the unreachable set $R(Q_{c'}, \emptyset)$ does not contain R_b and so by similar figures its furthest event from x is $R_{b'}$ where $[x, R_{b'}, R_b]$. Thus the two circular regions $R(Q_a, \emptyset)$ and $R(Q_{c'}, \emptyset)$ have an empty intersection from which we see that Q_c is the closest (real positive) event of $Q(Q_a, R, x, \emptyset)$ to x. Then as in the usual real Minkowski space–time

$$c = \frac{1 - v}{1 + v} \cdot a \quad (v \text{ positive real}) .$$

If we next consider a phase transformation of the Z_1–coordinate (that is, a coordinate transformation — not an interval–preserving bijection of M_{O5}) we see that the centres and radii of the unreachable sets are unchanged, although the phase of the relative velocity is changed, so now

$$c = \frac{1 - |v|}{1 + |v|} \cdot a \ . \tag{2a}$$

Other coordinate transformations which leave the origin and Q_a invariant are obtained from the product of all phase transformations and all orthogonal transformations in the $Z_1 Z_2 Z_3$–subspace (as in the usual real Minkowski space–time), so we see that $(2a)$ applies in general with

$$|v| \ = (v_1 v_1^* + v_2 v_2^* + v_3 v_3^*)^{1/2} \ . \tag{2b}$$

As with the usual real standard model of Minkowski space–time, the group of isometries (which leave the events of Q invariant) is isomorphic to the group of coordinate transformations (just described). Thus for any two paths R, S as in the statement of preconditions for the Axiom of Isotropy, there is an isometry which satisfies the full statement of the axiom.

The Axiom of Continuity   This axiom is obviously satisfied for the set of points on any line in a complex plane.

11.6 Independence model M_{O6}

The independence model M_{O6} is de Sitter space–time in which antipodal points are identified.

The de Sitter universe   We first describe the de Sitter universe (without antipodal identification). It can be described globally in terms of the single–shell (hyper–)hyperboloid

$$x^2 + u^2 + v^2 + y^2 - z^2 = 1 \tag{$\mathcal{H}_4$}$$

in $\mathbb{R}^5$ with the intrinsic indefinite metric induced on this hypersurface by the pseudo–Euclidean indefinite metric form

$$ds^2 = -dx^2 - du^2 - dv^2 - dy^2 + dz^2$$

in $\mathbb{R}^5$. This model of the de Sitter universe is described in considerable detail in the elegant monograph of Schrödinger (1956) who shows that the universe is a space–time of constant curvature, that its geodesics are generated by intersections with planes through the origin of $\mathbb{R}^5$, and that its linear automophisms are induced by the homogeneous Lorentz group in $\mathbb{R}^5$.

If $\mathbb{R}^5$ is equipped also with the obvious Euclidean metric, then the planes through the origin which make an angle less than 45^0 with the Z–axis intersect the hyper–hyperboloid in timelike geodesics and, similarly, planes inclined at angles greater than or equal to 45^0 with the Z–axis intersect the hyper–hyperboloid in spacelike and null geodesics, respectively. These geodesics are plane conic sections: the timelike geodesics are branches of hyperbolas, the spacelike geodesics are ellipses, and the null geodesics are straight lines. The "reduced model" is easily visualised in the XYZ–subspace as the hyperboloid

$$x^2 + y^2 - z^2 = 1 . \tag{$\mathcal{H}_2$}$$

The independence model M_{O6}   The independence model M_{O6} is the elliptic interpretation of de Sitter space–time. It is obtained by identifying antipodal pairs of points and regarding the pair as an event: thus in $\mathcal{H}_4$, an event is a pair of points $\{(x, u, v, y, z), (-x, -u, -v, -y, -z)\}$ and in $\mathcal{H}_2$ an event is a pair of points $\{(x, y, z), (-x, -y, -z)\}$. With the antipodal identification, de Sitter space–time is still an interesting cosmological model and is discussed in some detail by Schrödinger (1956).

The elliptic interpretation may also be described by projecting from the point $(0, 0, 0, 0, 0)$ onto the hyperplane (P^4) at infinity in P^5. Then the events are points exterior to the non–degenerate non–ruled quadric (ellipsoid)

$$x_1^2 + x_2^2 + x_3^2 + x_4^2 - x_5^2 = 0 \tag{$\mathcal{Q}$}$$

in the projective space P^4 and the paths are (the exterior segments of) straight lines which intersect the interior of the quadric.

The relation of betweenness is defined for triples of events on each path in the obvious way.

The axioms of order   Each path is order–isomorphic to an open subset of the real line, so Axioms O1–O5 obviously apply.

In this model the axiom of collinearity (Axiom O6) does not apply. The configuration $abcde$, of the statement of the axiom, defines a plane which intersects the quadric $\mathcal{Q}$ in an ellipse. The events are exterior to the ellipse and the paths are the exterior subsets of lines which meet the interior of the ellipse. Two counter examples to the axiom of collinearity (Axiom O6) are shown in Figure 92: one example does

204

not have an event f while the other example does have an event f but it is ordered such that $[baf]$.

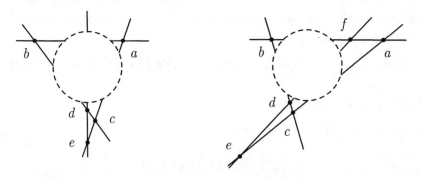

Figure 92

The axioms of incidence   If the model is considered as the set of points exterior to the quadric (ellipsoid) Q, it is easy to verify that all the Axioms of Incidence (I1–I7) are satisfied.

The Axiom of Isotropy   The proof of isotropy for the model M_{O4} can be modified so as to apply to model M_{O6}. Consider the lines in P^4 which contain the paths Q, R, S as segments and, as for model M_{O4} define the polar plane Π and points of intersection q, r, s as the intersections of these lines with the polar plane Π. Now the proof of isotropy given for model M_{O4} applies without any further modification.

The Axiom of Continuity   This axiom (Axiom C) is satisfied since each path is order–isomorphic to the reals.

11.7 Independence model M_{I1}

For the independence model M_{I1} we take $\mathcal{E}$ to be the empty set, $\mathcal{E} = \emptyset$, which has the single subset, $Q = \emptyset$, so we define the set of paths to be $\mathcal{P} = \{Q\} = \{\emptyset\}$ and we define $[abc]$ to mean that a, b, c are distinct. Thus

$$M_{I1} = \langle \emptyset, \{\emptyset\}, [\cdots] \rangle .$$

11.8 Independence model M_{I2}

The model M_{I2} consists of two copies of the usual cartesian model for Mikowski space–time which have no common event. Isotropy mappings are defined to act in the usual way on the copy which contains the event of coincidence, while leaving all events of the other copy invariant.

All axioms except the Axiom of Connectedness (Axiom I2) are satisfied.

11.9 Independence model M_{I3}

This model consists of ten events and four paths

$$Q^{(1)} = \{0, (1,1), (1,2), 3\}$$

$$Q^{(2)} = \{0, (2,1), (2,2), 3\}$$

$$Q^{(3)} = \{0, (3,1), (3,2), 3\}$$

$$Q^{(4)} = \{0, (4,1), (4,2), 3\}$$

with the relation of betweenness defined in the obvious way on each path. The model M_{I3} is illustrated in Figure 93.

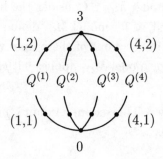

Figure 93

All paths meet at the events 0 and 3, so the model does not satisfy the Axiom of Uniqueness of Paths (Axiom I3).

All the axioms of order are satisfied, with Axiom O6 being satisfied trivially. All axioms of incidence, with the exception of Axiom I3, are satisfied.

The given conditions for the statement of the Axiom of Isotropy are satisfied for any three distinct paths $Q^{(i)}, Q^{(j)}, Q^{(k)}$ taking the place of Q, R, S and with either event 0 (resp. event 3) taking the place of event x, and with $(i, 2)$ (resp. $(i, 1)$) taking the place of the event Q_a. It is easily verified that the permutation which interchanges the paths $Q^{(j)}$ and $Q^{(k)}$ is an isotropy mapping.

Finally the Axiom of Continuity is satisfied trivially.

11.10 Independence models M_{I4} and M'_{I4}

If the usual cartesian model of Minkowski space–time is referred to as a $3 + 1$–dimensional model then suitable models for M_{I4} are the corresponding $1 + 1$–dimensional or $4 + 1$–dimensional models.

11.11 Independence model M_{I5}

Galilean space–time is the cartesian product $\mathbb{R} \times \mathrm{E}^3$ of the reals with three–dimensional Euclidean geometry, where straight lines with finite velocity are the paths and $[\cdots]$ is defined on each path to represent the usual betweenness relation. The set of isotropy mappings includes expansions and contractions, as well as rotations and reflections, in the three–dimensional Euclidean space.

Apart from Axiom I5, Galilean space–time satisfies all the remaining axioms.

11.12 Independence model M_{I6}

The field Ω^*   Let Ω^* be a subfield of the reals whose definition resembles that of the field Ω described by Hilbert (1913) to demonstrate the independence of the Axiom of Continuity in Euclidean geometry. The field Ω^* is the set of all algebraic numbers that arise from the number 1 and the application a finite number of times of the four arithmetic operations of addition, subtraction, multiplication, division and the fifth operation $\sqrt{1 + \omega^2}$, together with the sixth operation $\sqrt{1 - \omega^2}$, for numbers $\omega \in \Omega^*$ such that $1 - \omega^2 \geq 0$ where in all these expressions the number ω denotes a number which arises from these operations.

The reason for including the sixth operation is to permit the existence of expressions of the form $\sqrt{1 - \omega^2}$ which occur in the Lorentz transformation formulae.

The model M_{I6}   The model M_{I6} has the same set of events $\mathcal{E}$ and the same betweenness relation as for the usual cartesian model of Minkowski space–time with the usual invariant squared interval. The set of paths $\mathcal{P}$ is the subset of the usual set of timelike lines whose 3–velocities belong to $\Omega^* \times \Omega^* \times \Omega^*$. Thus Axiom I6 is not satisfied.

The Axioms of Order O1–O5 are obviously satisfied. Axiom O6 is satisfied because Ω^* is a field. The Axiom of Dimension (Axiom I4) is also satisfied because Ω^* is a field, and it is readily seen that the remaining axioms of incidence are satisfied. The group of linear isometries is analogous to the Lorentz group except that it is defined over the field Ω^*, instead of over the field $\mathbb{R}$, so the model M_{I6} satisfies the Axiom of Isotropy (Axiom S). Finally the Axiom of Continuity (Axiom C) is satisfied since each path is order–isomorphic to the reals.

11.13 Independence model M_{I7}

This model is defined in a similar way to the model M_{I3}. The model M_{I7} consists of nine events and four paths

$$Q^{(1)} = \{0, (1, 1), (1, 2)\}$$

$$Q^{(2)} = \{0, (2, 1), (2, 2)\}$$

$$Q^{(3)} = \{0, (3, 1), (3, 2)\}$$

$$Q^{(4)} = \{0, (4, 1), (4, 2)\}$$

with the obvious relations of betweenness for triples of events on each path. The model does not satisfy Axiom I7. All other axioms are satisfied for reasons similar to those given for the independence model M_{I3}.

11.14 Independence model M_S

This model resembles the usual cartesian model of Minkowski space–time except that the timelike lines have 3–velocities which lie within a "non–quadric" cone

$$|v_1| + |v_2| + |v_3| > 1 .$$

The Axiom of Isotropy (Axiom S) is not generally satisfied. All other axioms are satisfied.

11.15 Independence models M_{C_1} , M_{C_2} and M_{C_3}

For Euclidean geometry, Hilbert (1913) states two "Axioms of Continuity" and provides independence models for both. In the present axiomatic system there is a single axiom of continuity but we describe three independence models. Two models resemble those provided by Hilbert and these models resemble Minkowski space–time in their symmetries and kinematic relations, while the third model has a closer resemblance to Galilean space–time rather than Minkowski space–time. The first model M_{C_1} is based on the subfield Ω^* of the reals, while the second and third models M_{C_2} and M_{C_3} are based on a ring of formal power series $\mathbb{R}(\varepsilon)$ defined over the reals.

The model M_{C_1} This model is specified in the same way as the usual cartesian model of Minkowski space–time except that it is defined over $\Omega^* \times \Omega^* \times \Omega^* \times \Omega^*$ instead of over $\mathbb{R}^4$. (Before examining this model, the reader is encouraged to peruse the definition of the subfield Ω^* which is discussed in detail in Section 11.12 before the description of the independence model M_{I6}).

In the model M_{C_1} the axioms of order are satisfied as in the model M_{I6}. In the present model, Axiom I6 is also satisfied. The other Axioms of Incidence, as well as the Axiom of Isotropy (Axiom S), are satisfied for the same reasons as for the model M_{I6}. Finally, as with the field Ω described by Hilbert (1913), the field Ω^* does not satisfy the Axiom of Continuity (Axiom C), so the sets of events of the paths of M_{C_1} do not satisfy the axiom.

Alternative independence model M_{C_2} The previous independence model could be regarded as a subset of Minkowski space–time. Another model, which contains Minkowski space–time as a subset, can be obtained by considering the ring of formal power series in one variable instead of the field of real numbers. This model resembles a model $\Omega(t)$ used by Hilbert (1913) to demonstrate the independence of "Archimedes Axiom" (the first axiom of the group of two axioms of continuity stated by Hilbert).

The ring of formal power series Let $\mathbb{R}$ denote the reals and let $\mathbb{R}(\varepsilon)$ denote the *ring of formal power series* in the variable ε. Thus

$$\mathbb{R}(\varepsilon) = \left\{ \sum_{n=0}^{\infty} a_n \varepsilon^n : a_n \in \mathbb{R} \right\}$$

where addition and multiplication are defined as usual for power series and where the binomial theorem is considered to apply without any convergence requirements. The ring $\mathbb{R}(\varepsilon)$ is not a field since there are elements, such as

$$0 + 1\varepsilon^1 + 0 + 0 + \ldots$$

which have *no multiplicative inverses*. Elements of $\mathbb{R}(\varepsilon)$ which have coefficients that are zero for $n \geq 1$ form a subfield which is isomorphic to the reals.

A *linear ordering* on the elements of $\mathbb{R}(\varepsilon)$ can be defined in the following way: let $(a_n : n = 0, 1, 2, \cdots)$ and $(b_n : n = 0, 1, 2, \cdots)$ be the sequences of real coefficients of the elements a and b respectively. Then we say that

$$a \prec b \iff \exists \text{ an integer n such that } \begin{cases} \text{(i) } \forall i < n, \ a_i = b_i \\ \text{(ii) } a_n < b_n \ . \end{cases}$$

It is easily verified that this relation is irreflexive, asymmetric and transitive and that for any two distinct elements a and b, either $a \prec b$ or $b \prec a$. If two elements differ in their first (ie. subscript 0) coordinates then we use the usual inequality symbol $<$ (and write $a < b$ or $a > b$ as the case may be). (Thus the familiar inequality symbol occurring in the relation $a < b$ implies the new relation $a \prec b$, but not vice versa).

We now define the independence model M_{C_2} in much the same way as the usual Minkowski space–time; namely

$$M_{O6} = \mathbb{R}(\varepsilon) \times \mathbb{R}(\varepsilon) \times \mathbb{R}(\varepsilon) \times \mathbb{R}(\varepsilon)$$

with the usual inner product

$$u \cdot v = u_0 v_0 - u_1 v_1 - u_2 v_2 - u_3 v_3$$

and the usual criteria for straight lines. However there is a difference in the definition of timelike lines and timelike vectors where we are obliged to use the $<$ inequality: accordingly we say that a vector u is a *timelike vector* if and only if $u \cdot u > 0$. For this model there are automorphisms analogous to the Lorentz transformations and the restrictions on the coefficients of the matrices representing them have exactly the same form as for the Lorentz transformations, where the usual $<$ inequality applies in both models. The relation of betweenness is defined as in the usual cartesian model for Minkowski space–time.

All axioms of incidence are satisfied — the Axiom of Uniqueness (Axiom I3) is satisfied since $\mathbb{R}(\varepsilon)$ is an integral domain. The Axiom of Symmetry (Axiom S) is satisfied due to the definition of $\mathbb{R}(\varepsilon)$. However the Axiom of Continuity (Axiom C) is not satisfied for the model M_{C_2} as may be seen by considering the bounded sequence of elements of $\mathbb{R}(\varepsilon)$

$$e_{1n} = 1 + n(\varepsilon^1 + \varepsilon^2 + \varepsilon^3 + \cdots)$$

which has no least upper bound and which can be used to define a bounded chain of events (on a path) with no closest bound.

Clearly the axioms of order (O1–O5) are satisfied. To verify the Axiom of Collinearity (Axiom O6) we first observe that the analogues of Lorentz transformations map collinear sets onto collinear sets, so we need only verify the incidence property of the axiom. We first consider the special case of two paths in the TX–subspace (ie. in the X_0X_1–subspace) and show that, if they cross, they must also meet. Consider paths with equations $x = wt$ and $x = -v(t - t_0)$: the condition that they meet is that

$$wt = -v(t - t_0)$$

which is satisfied only if

$$t = \frac{v}{v + w} \cdot t_0$$

and it is easily verified that this equation always has a solution for $v, w \succ 0$. For the general case, where two paths cross in a more general collinear set, a translation followed by an orthogonal transformation and then a Lorentz transformation will map the collinear set onto the TX–subspace already considered.

Alternative independence model M_{C_3} An independence model M_{C_3} can also be defined over the cartesian product

$$\mathbb{R}(\varepsilon) \times \mathbb{R}(\varepsilon) \times \mathbb{R}(\varepsilon) \times \mathbb{R}(\varepsilon)$$

by taking an analogue of Galilean space time, where the set of paths is the set of lines with 4–velocity vectors (v_0, v_1, v_2, v_3) such that $v_0 > 0$ (in the sense of $>$ rather than $\succ$ as defined for the ring of formal power series $\mathbb{R}(\varepsilon)$ above). This model bears a stronger resemblance to Galilean space time, rather than Minkowski space time, in its symmetries and kinematic relations. The set of timelike lines has no upper bound for velocities. Like affine geometry, model M_{C_3} has no obvious metric.

Appendix 1
Veblen's axioms for ordered geometry

The name "ordered geometry" has been used by Coxeter (1961) to describe the "geometry of serial order" which can be developed in terms of a single undefined ternary relation of order, called "betweenness" or "intermediacy". This geometry can be developed without any mention of congruence and without any axiom of uniqueness of parallelism. Two approaches to this geometry are given by Coxeter (1965) and Veblen (1904). The discussion of Coxeter is based on that of Veblen but is more detailed. Veblen (1904) actually states a total of twelve axioms – the twelfth axiom is not required for "ordered geometry" but is stated by Veblen in order to discuss affine geometry after first developing ordered geometry. The first eleven axioms of Veblen (1904), which are cited in Theorem 80 of Section 9.1, are reproduced[1] below:

AXIOM I. There exist at least two distinct points.

AXIOM II. If points A, B, C are in the order ABC, they are in the order CBA.

AXIOM III. If points A, B, C are in the order ABC, they are not in the order BCA.

AXIOM IV. If points A, B, C are in the order ABC, then A is distinct from C.

AXIOM V. If A and B are any two distinct points, there exists a point C such that A, B, C are in the order ABC.

DEF. 1. The *line* $AB(A \neq B)$ consists of A and B and all points X in one of the possible orders ABX, AXB, XAB. The points X in the order AXB constitute the *segment* AB. A and B are the *end-points* of the segment.

AXIOM VI. If points C and D $(C \neq D)$ lie on the line AB, then A lies on the line CD.

AXIOM VII. If there exist three distinct points, there exist three points A, B, C not in any of the orders ABC, BCA, or CAB.

DEF. 2.   Three distinct points not lying on the same line are the *vertices* of a *triangle* ABC, whose *sides* are the segments AB, BC, CA, and whose *boundary* consists of its vertices and the points of its sides.

AXIOM VIII.   If three distinct points $A, B,$ and C do not lie on the same line, and D and E are two points in the orders BCD and CEA, then a point F exists in the order AFB and such that D, E, F lie on the same line.

DEF. 5.   [sic][2]. A point O is *in the interior of* a triangle if it lies on a segment, the end-points of which are points of different sides of the triangle. The set of such points O is *the interior* of the triangle.

DEF. 6.   If A, B, C form a triangle, the *plane ABC* consists of all points collinear with any two points of the sides of the triangle.

AXIOM IX.   If there exist three points not lying in the same line, there exists a plane ABC such that there is a point D not lying in the plane ABC.

DEF. 7.   If $A, B, C,$ and D are four points not lying in the same plane, they form a *tetrahedron $ABCD$* whose *faces* are the interiors of the triangles ABC, BCD, CDA, DAB (if the triangles exist) whose *vertices* are the four points, $A, B, C,$ and D, and whose *edges* are the segments AB, BC, CD, DA, AC, BD. The points of faces, edges, and vertices constitute the *surface* of the tetrahedron.

DEF. 8.   If A, B, C, D are the vertices of a tetrahedron, the space $ABCD$ consists of all points collinear with any two points of the faces of the tetrahedron.

AXIOM X.   If there exist four points neither lying in the same line nor lying in the same plane, there exists a space $ABCD$ such that there is no point E not collinear with two points of the space, $ABCD$.

AXIOM XI.   If there exists an infinitude of points, there exists a certain pair of points AC such that if $[\sigma]^*$ is any infinite set of segments of the line AC, having the property that each point which is A, C or a point of the segment AC is a point of a segment σ, then there is a finite subset $\sigma_1, \sigma_2, \cdots, \sigma_n$ with the same property.

AXIOM XII.   If a is any line of any plane α there is some point C of α through which there is not more than one line of the plane α which does not intersect a.

(The notation $[\sigma]$ used in the statement of Axiom XI is explained by Veblen in a footnote as " * $[e]$ denotes a set or class of elements, any one of which is denoted by e alone or with an index or subscript." In a footnote to Axiom II, Veblen states that the axiom is not intended to imply that A, B, C are distinct).

Veblen shows that the *first eleven axioms imply that the set of points is an open convex (three–dimensional) subset of three–dimensional projective space.* This is the result to which reference is made in Sections 9.1 and 9.2. The twelfth axiom of Veblen does not apply to the situation which we are considering. It is used by Veblen to provide an independent system of axioms for affine geometry. Thus *Veblen's axioms I–XI are axioms for ordered geometry* while his axioms I–XII are axioms for affine geometry.

Appendix 2
Alternative axiom systems

The system of axioms stated in Chapter 2 is a system of independent axioms. The primitive undefined basis and some of the axioms were expressed in a form which would allow the existence of independence models for all axioms: as a result the expression of some of the axioms is slightly complicated and the intuitive meaning and significance is somewhat obscured. To clarify the meaning of the axioms and to state them in a more directly usable form, an alternative set of axioms is stated below.

Some of these axioms are stated in exactly the same way as the corresponding axioms of the independent system (namely Axioms O1–O4, I1–I5, S and C). Axioms of the alternative system whose statements differ from those of the independent system are indicated with asterisks as superscripts (namely Axioms O5,O6,I6,I7)[1].

These axioms lead to the same theorems as the previous set of axioms but they do not permit the existence of some of the independence models.

The reader is encouraged to read the introductory comments to Chapter 2 which also apply to this appendix.

A2.1 Primitive undefined basis

Minkowski space time is

$$\mathcal{M}^* = \langle \mathcal{E}, \mathcal{P}, [\cdots] \rangle$$

where $\mathcal{E}$ is a set whose elements are called *events*, $\mathcal{P}$ is a set of subsets of $\mathcal{E}$ called *paths* and $[\cdots]$ is a ternary relation on the set of events of $\mathcal{E}$ called a *betweenness relation*.

Paths will be denoted by upper case symbols $Q, R, S, \cdots$; events will be denoted by lower case symbols $a, b, c, d, \cdots$ or by a path symbol with a subscript, such as for example Q_a, Q_1, Q_x, Q_α for events which belong to the path Q and $R_b, R_3, R_w R_\gamma$ for events which belong to the path R. Given a pair of distinct events $a, b \in S$ we say that a, b *belong to* S or *lie on* S or that they can *be connected by* S or that S *passes through* them.

We will presuppose logic, set theory and the arithmetic of real numbers.

A2.2 Axioms of order

The axioms of order resemble axioms or theorems of the geometric axiom systems of Hilbert (1899, 1913), Veblen (1904, 1911), Moore (1908) and Veblen and Young (1908) for Euclidean and hyperbolic geometry. The first five axioms of order (Axioms O1–O4, O5*) ensure that each path is a linearly ordered set of events.

Axiom O1
For events $a, b, c \in \mathcal{E}$,

$$[abc] \implies \exists Q \in \mathcal{P} : \ a, b, c \in Q .$$

Axiom O2
For events $a, b, c \in \mathcal{E}$,

$$[abc] \implies [cba] .$$

Axiom O3
For events $a, b, c \in \mathcal{E}$,

$$[abc] \implies a, b, c \ \text{are distinct} .$$

Axiom O4
For distinct events $a, b, c, d \in \mathcal{E}$,

$$[abc] \ \text{and} \ [bcd] \implies [abd] .$$

Axiom O5
For any path $Q \in \mathcal{P}$ and any three distinct events $a, b, c \in Q$,

$$[abc] \quad or \quad [bca] \quad or \quad [cab].$$

The axiom of collinearity (Axiom O6*)   This axiom is a kinematic analogue of the geometric axiom of plane order given by Veblen (1904, 1911) and Moore (1908) and makes it possible to discuss "rectilinear motion" in terms of "collinear sets" of events and paths. To obtain an appreciation of the kinematic ideas expressed by the axiom of collinearity, the reader is encouraged to draw a diagram corresponding to the statement of the axiom. If a template is made by cutting a narrow slit in a sheet of paper, the paths may be observed "in motion" by moving the template gradually across the diagram. This axiom is an analogue of the geometric "Axiom of Pasch" as stated by Veblen (1904) and Moore (1908).

Axiom O6*

If Q, R, S are distinct paths which meet at events $a \in Q \cap R$, $b \in Q \cap S$, $c \in R \cap S$ and if:

(i) *there is an event $d \in S$ such that $[bcd]$, and*

(ii) *there is an event $e \in R$ and a path T which passes through both d and e such that $[cea]$,*

then T meets Q in an event f such that $[afb]$.

As is the case with the analogous geometric axiom of plane order, this axiom makes statements about both incidence and order. The ensueing development of "order" properties on paths and properties of "collinear sets" has much in common with the treatment of order properties of points on lines and properties of planes as given by Veblen (1904, 1911). However the investigation of space–time is more complicated due to the existence of pairs of events which can not be connected by paths (see Axiom I5). As a consequence, the existence of "collinear sets" can only be established at the conclusion of Chapter 5 and is based upon thirty five preceding theorems.

A2.3 Axioms of incidence

Of the seven axioms of incidence which follow, the first four have counterparts in the axiomatic systems of Hilbert (1899), Veblen (1904, 1911) and Moore (1908) for Euclidean geometry. The fifth, sixth and seventh axioms describe properties of the "unreachable set" which distinguishes a space–time from a geometry. These latter properties are essentially properties of "non-incidence" and involve both order properties and incidence properties in their formulations. Axiom I5 excludes Galilean space–time as a possible model, while Axiom I6* and Axiom I7* specify the "connectedness" and "boundedness" of the unreachable set.

Axiom I1 (Existence)
$\mathcal{E}$ is not empty.

Axiom I2 (Connectedness)
For any two distinct events $a, b \in \mathcal{E}$ there are paths R, S such that $a \in R$, $b \in S$ and $R \cap S \neq \emptyset$.

Axiom I3 (Uniqueness)
For any two distinct events, there is at most one path which contains both of them.

In the subsequent development we will frequently be discussing the properties of sets of paths which meet at a given event. We will call any such set a *SPRAY of paths*, or more concisely a *SPRAY*, where the upper case letters indicate that we are referring to a set of paths rather than to a set of events: given any event x, we define

$$SPR[x] := \{R: \ R \ni x, \ R \in \mathcal{P}\} \ .$$

The corresponding set of events is called a *spray*, where the lower case letters indicate a set of events. We define

$$spr[x] := \{R_y: \ R_y \in R, \ R \in SPR[x]\} \ .$$

A subset of three paths $\{Q, R, S\}$ of a SPRAY is *dependent* if there is a path which does not belong to the SPRAY and which contains one event from each of the three paths: we also say that any one of the three paths is *dependent on* the other two. Otherwise the subset is *independent*. We next give recursive definitions of dependence and independence which will be used to characterize the concept of dimension. A path T is *dependent on* the set of n paths (where $n \geq 3$)

$$S = \{Q^{(i)}: \ i = 1, 2, \ldots, n; \ Q^{(i)} \in SPR[x]\}$$

if it is dependent on two paths $S^{(1)}$ and $S^{(2)}$, where each of these two paths is dependent on some subset of $n - 1$ paths from the set S. We also say that the set of $n + 1$ paths $S \cup \{T\}$ is a *dependent set*. If the set of $n + 1$ paths has no dependent subset, we say that the set of $n + 1$ paths is an *independent set*.

We now make the following definition which enables us to specify the dimension of Minkowski space–time. A SPRAY is a *3-SPRAY* if:

(i)   it contains four independent paths, and

(ii) all paths of the SPRAY are dependent on these four paths.

Axiom I4 (Dimension)
If $\mathcal{E}$ is non–empty, then there is at least one 3-SPRAY.

Given a path Q and an event $b \notin Q$, we define the *unreachable subset of Q from b* to be

$$Q(b, \emptyset) := \{x: \ \text{there is no path which contains } b \text{ and } x, \ x \in Q\} \ .$$

That is, the unreachable subset of Q from b is the subset of events of Q which can not be joined to b by a single path. If two events can not be connected by any path, we say that each is *unreachable* from the other; otherwise each is *reachable* from the other.

Axiom I5 (Non–Galilean)

For any path Q and any event $b \notin Q$, the unreachable set $Q(b, \emptyset)$ contains (at least) two events.

Axiom I6* (Connectedness of the Unreachable Set)

Given any path Q, any event $b \notin Q$, and distinct events $Q_x, Q_z \in Q(b, \emptyset)$, then for all $Q_y \in Q$,

$$[\, Q_x \, Q_y \, Q_z \,] \Longrightarrow Q_y \in Q(b, \emptyset) \ .$$

Axiom I7* (Boundedness of the Unreachable Set)

Given any path Q, any event $b \notin Q$ and events $Q_x \in Q \setminus Q(b, \emptyset)$ and $Q_y \in Q(b, \emptyset)$, there is an event $Q_z \in Q \setminus Q(b, \emptyset)$ such that

$$[\, Q_x \, Q_y \, Q_z \,] \ .$$

A2.4 Axiom of isotropy or symmetry

Compared to the absolute geometries, Minkowski space–time has the additional structure provided by the existence and properties of unreachable sets (Axioms I5, I6*, I7*). These properties, together with the single property of isotropy of the following axiom, are sufficient to take the place of all the axioms of congruence and the axiom of uniqueness of parallels used by Hilbert (1899), Moore (1908) and Veblen (1911) for Euclidean geometry.

For any two distinct paths Q, R which meet at an event x, we define the *unreachable subset of Q from Q_a via R* to be

$$Q(Q_a, R, x, \emptyset) := \{Q_y : [x \, Q_y \, Q_a] \text{ and } \exists R_w \in R \text{ such that } Q_a, Q_y \in Q(R_w, \emptyset) \} \ .$$

The statement of the Axiom of Isotropy can be expressed intuitively as: "if an observer Q observes that paths R and S appear to move in the same way, then there is an automorphism of $\mathcal{M}^*$ which leaves the events of Q invariant and maps R onto S".

Axiom S (Symmetry or Isotropy)

If Q, R, S are distinct paths which meet at some event x and if $Q_a \in Q$ is an event distinct from x such that

$$Q(Q_a, R, x, \emptyset) = Q(Q_a, S, x, \emptyset)$$

then

    (i)   *there is a mapping $\theta : \mathcal{E} \longrightarrow \mathcal{E}$*

    (ii) *which induces a bijection $\Theta : \mathcal{P} \longrightarrow \mathcal{P}$*

such that

    (iii) *the events of Q are invariant, and*

    (iv) $\Theta : R \longrightarrow S$.

The mapping θ is called a *symmetry mapping* or an *isotropy mapping*, with Q as the *invariant path* .

A2.5 Axiom of continuity

This final axiom resembles the second–order geometric axiom of the same name in the axiom systems of Hilbert (1899), Veblen (1904, 1911) and Moore (1908).

Before we state the axiom we make the definition: a sequence of events

$$Q_x = Q_0, \; Q_1, \; Q_2, \; \cdots$$

(of a path Q) is called a *chain* if:

    (i)   it has two distinct events, or

    (ii) it has more than two distinct events and for each Q_i other than Q_0 and Q_1,

$$[Q_{i-2} \; Q_{i-1} \; Q_i] \; .$$

A *finite chain* is denoted by writing $[Q_0 \; Q_1 \; Q_2 \; \cdots \; Q_n]$ and an *infinite chain* is denoted by writing $[Q_0 \; Q_1 \; Q_2 \; \cdots]$ (note that the concept of a "chain" used by Veblen and Young (1908) for the discussion of projective geometries is entirely different from the concept defined above).

Given a path $Q \in \mathcal{P}$ and an infinite chain $[\, Q_0 \; Q_1 \; \cdots \;]$ of events in Q, the set

$$\mathcal{B} = \{Q_b : \; \forall Q_i \neq Q_0, \; [Q_0 \; Q_i \; Q_b]; \; Q_b \in Q\}$$

is called the *set of bounds*[2] of the chain: if $\mathcal{B}$ is non–empty we say that the chain is *bounded*. If there is a bound $Q_b \in \mathcal{B}$ such that for all $Q_{b'} \in \mathcal{B} \setminus \{Q_b\}$,

$$[Q_0 \; Q_b \; Q_{b'}]$$

we say that Q_b is a *closest bound*.

Axiom C* (Continuity)
Any bounded infinite chain has a closest bound.

Appendix 3
Open questions

There are several interesting issues related to the exacting requirements of independence and form of the set of axioms and to the structure of independence models. Some of these aspects are discussed in the following Appendix 4 "Notes to the Chapters" (pp. 225–230). A number of open questions have arisen and are described here as problems for future research.

Problem 1:   Alternative statement for Axiom O6

Axiom O6 makes two statements: one concerning incidence, the other concerning order. The independence model M_{O6} violates both properties. Could Axiom O6 be replaced by a weaker axiom?

Problem 2:   Alternative statement for Axiom S

Is it possible to restate the Axiom of Isotropy so that isotropy mappings preserve the relationship of betweennness, and still have an independent set of axioms? The other axioms would require some modifications.

Problem 3:   Alternative independence model for Axiom O4

Our axiomatic system specifies Minkowski space–time as

$$\mathcal{M} = \langle \mathcal{E}, \mathcal{P}, [\cdots] \rangle$$

and the isotropy mappings preserve the relation of betweenness for $\mathcal{M}$, so they are actually automorphisms of $\mathcal{M}$.

A disadvantage of the independence models M'_{O4} and M'_{O4} is that, in general, their isotropy mappings do not preserve the relations of betweenness, so that isotropy mappings are not automorphisms. The axiomatic system described in Chapter 2 does not state that isotropy mappings are automorphisms of (the full structure of) $\mathcal{M}$, it merely states that isotropy mappings are automorphisms of the reduced structure $\langle \mathcal{E}, \mathcal{P} \rangle$. However for Minkowski space–time and for all the other independence models, isotropy mappings are automorphisms. It would be desirable to find an independence model for Axiom O4 in which isotropy mappings are automorphisms. If this could be done the statement of Axiom S could be amended so that isotropy mappings are stated to be automorphisms[1] of $\mathcal{M}$

A useful project would be to find an independence model for Axiom O4 in which isotropy mappings are automorphisms or else to show that there can be no such model.

Another alternative would be to specify Minkowski space–time using a four–event relation of "separation" comparable to that used for projective geometry. This relation would, of course, be preserved by the isotropy mappings of models M_{O4} and M'_{O4} since the isotropy mappings are projectivities.

Problem 4: Alternative independence models for Axiom O6

For Euclidean geometry the axiom analogous to our Axiom O6 is the Axiom of Pasch which is stated in various forms in different axiomatic systems. The significance of the Axiom of Pasch and properties of the corresponding independence models, which are known as "Pasch–free" Euclidean geometries, are discussed by Schnabel and Pambuccian (1985), Szczerba (1970), Szczerba and Szmielew (1970) and Szmielew (1980).

Is it possible to find alternative independence models for Axiom O6?

Problem 5: First order axiomatization

In order to consider questions of consistency and decidability alternative first–order theories have been considered by some authors. First order axiom schema to replace the Axiom of Continuity have been discussed by Tarski (1951, 1959, 1967) for geometries. Goldblatt (1987) has specified Minkowski space–time in a first–order theory with a 3-term predicate of betweenness and a 4-term predicate of orthogonality. These theories are complete and decidable.

If the present Axiom of Continuity is replaced by an axiom schema of all its first order instances, the resulting axiomatic system has many models. Three of these models are the standard coordinate model M (as described in Chapter 10) and M_{C_2} and M_{C_3} (as described in Chapter 11). What are the other models? Would the system of axioms be sufficient to prove some sort of "causality theorem"[2,3] resembling Theorem 26?

Whereas the standard model M and the models M_{C_1} and M_{C_2} have symmetries and metric relations whose expressions become indistinguishable in the physical sense of coping with experimental error, the model M_{C_3} is essentially different from the previous three models since it has symmetries which bear a closer resemblance to Galilean space–time and, like affine geometry, has no metric.

Thus we can not establish a representation theorem analogous to those of Tarski and Goldblatt. What modifications would be required to prove the existence of a metric? Would the modified system of axioms and continuity schema be mutually independent?

Problem 6:   Independent axioms for Euclidean geometry

Veblen (1904) has given a system of independent axioms for affine geometry and Moore (1908) modified this system to obtain a system of independent axioms for Euclidean geometry stated in terms of the undefined primitive elements called points and two undefined relations of betweenness and congruence. In the axiomatic system of Moore (1908), lines were defined using the relation of betweenness. This approach eliminates the need for an axiom which would correspond to (part of) the first proposition of Axiom II,1 of Hilbert (1899) or to Axiom O1 of the present axiomatic system for Minkowski space–time. In Hilbert's system Axiom II,1 corresponds to the conjunction of our Axioms O1 and O2.

Is it possible to obtain a system of independent axioms for Euclidean geometry in terms of undefined "points", "lines" and relations of betweenness and congruence? That is, is it possible to state a categorical system of independent axioms for Euclidean geometry[4] which includes independent axioms analogous to Axiom O1 and Axiom O2 of the present axiomatic system for Minkowski space–time?

Problem 7:   Independent axioms for de–Sitter space–times

The de–Sitter space–times[5] and their covering spaces are homogeneous and isotropic. They have a property of non–uniqueness of parallelism analogous to Bolyai-Lobachevskian geometry. Is it possible to obtain systems of independent, categorical and consistent axioms for these space–times?

Problem 8:   Three–dimensional theorems

Is it possible to obtain the standard model of Minkowski space–time more directly using the uniqueness of parallelism (Theorem 68) and the affine invariance of time scales between intersecting collinear sets (Theorem 65(i))?

Appendix 4
Notes to the chapters

Notes to Chapter 1: Introduction

Note 1   A theory is said to be *categorical* or *categoric* if all its models are isomorphic.

Note 2   But see the notes to Chapter 2 where some differences of axiomatic style are discussed.

Notes to Chapter 2 : The axioms

Note 1   It is also possible to consider other alternative primitive undefined bases as discussed in notes 2,3 and 4 below.

Note 2   One alternative basis is to have "local betweennness" relations defined such that each path has its own betweenness relation which may be thought of intuitively as the subjective concept of a freely–moving inertial "observer" ascribing the notion of betweennness to events in his own "local" history.

The set of all such "betweennness relations" is, of course, an infinite set so the corresponding undefined primitive basis has an infinite set of undefined relations. With this primitive basis there is no need for an axiom corresponding to Axiom O1 of the present axiomatic system, so the resulting axiomatic system would have one less axiom but an infinite set of undefined betweenness relations.

Note 3   It is possible to dispense with $\mathcal{P}$ as a primitive concept by defining[4] $\mathcal{M} = \langle \mathcal{E}, [\cdots] \rangle$, where the set of paths $\mathcal{P}$ is then defined in much the same way as the set of lines (by authors such as Veblen (1904, 1911) and Moore (1908)) in the absolute geometries and $[\cdots]$ is a ternary betweenness relation on $\mathcal{E}$. Thus these authors define a line for each pair of distinct points a, b to be

$$ab := \{x : [xab], [axb], [abx], [bax], [bxa], [xba]\} \ \cup \{a, b\} \ .$$

The method of Veblen and Moore has the effect of including an analogue of the first proposition of Hilbert's Axiom II,1 (Hilbert (1899)) within the *definition* of a line.

If we compare the axiom systems of Veblen (1904, 1911) and Moore (1908) with that of Hilbert (1899) we see that Veblen and Moore's definition of a line obviates one of Hilbert's axiomatically stated postulates — namely the first proposition of Axiom II,1. This might appear to be an advantage but we also see that the possibility of a complex geometry is excluded in Hilbert's approach by the explicitly stated Axiom II,4, whereas in the systems of Veblen and Moore the very *definition* of a line *excludes* any possibility of a complex geometry even though their axiomatic systems have an axiom (Veblen 1904, Axiom VIII) which is very similar to Hilbert's Axiom II,4. The approaches of Veblen and Moore appear to dispense with the need for an assumption similar to Hilbert's Axiom II,1 (the first proposition only) by incorporating it in the definition of a line.

Note 4 The present axiomatic system for Minkowski space–time could be presented in a form $\mathcal{M} = \langle \mathcal{E}, [\cdots] \rangle$ similar to those of Veblen (1904, 1911) and Moore (1908) (as described in Note 3 above). The modified system would then have two less axioms (since the properties of Axioms O1 and O5 would be included in the definition of a path), the number of axioms would be thirteen, and the undefined set of paths $\mathcal{P}$ would no longer be required.

Why then, is this form of presentation not invoked here? The reasons are, firstly, the author prefers to introduce assumed properties within the statements of axioms rather than obscuring the assumptions by incorporating them within definitions. Secondly, by *defining* paths in terms of the betweenness relation, the possibility of complex space–times would be excluded by *definition*. Thirdly, for the present axiomatic system we are able to demonstrate independence models M_{O1} and M_{O5} for Axioms O1 and O5, respectively. It is not known whether this can be done for the corresponding modification of the axiomatic system of Moore (1908) for Euclidean geometry. (This remains an open question and is included in Appendix 3).

Note 5 The corresponding axiom of Hilbert (1899) is Axiom II,4 while Veblen (1904) states the stronger Axiom VIII which is equivalent to the statement of Theorem 3 of Chapter 3.

Note 6 The terms "mapping", "surjective" and "injective" are used according to Bourbaki (1968, Vol I, Chapter 2, Section 3).

Note 7 An alternative version of Axiom S can be stated with (i) and (ii) replaced by the statement that θ is an automorphism of $\mathcal{M} = \langle \mathcal{E}, \mathcal{P}, [\cdots] \rangle$ but this would replace the weaker properties (i) and (ii) with the stronger assumptions:

(i') $\theta : \mathcal{E} \longrightarrow \mathcal{E}$ is bijective, and

(ii') θ preserves betweenness relations.

This stronger axiom is not required since both properties (i') and (ii') are proved as theorems within the present axiom system. A further disadvantage of the stronger axiom is that it does not admit our models M_{O4} and $M_{O4'}$ as independence models for Axiom O4, since many of the isotropy mappings of these models do not preserve the relation of betweenness.

Notes to Chapter 5: Collinear sets

Note 1 $CSP\langle R, S \rangle$ is a linearly ordered set (by Theorem 33) so any of its member paths separates it into two "sides". A more detailed definition of the concept of left– and right–sides is given in the Crossing Theorem (Th.37) which follows the present theorem.

Notes to Chapter 9: Three–dimensional theorems

Note 1 Coxeter (1961, 1965) has used the name "ordered geometry" as a convenient abbreviation for "the geometry of serial order". For the purposes of our discussion, a *three–dimensional ordered geometry* is any collection of undefined elements called "points" which satisfy Axioms I–XI of Veblen (1904). (Axiom XII of Veblen is required for a discussion of affine geometry but is not needed and is not satisfied in the present context.) An alternative discussion which gives details complementing those provided by Veblen (1904) is given by Coxeter (1965) who provides further references. Both Veblen (1904) and Coxeter (1965) show that a three–dimensional ordered geometry can be embedded (as a convex three–dimensional subset) in three–dimensional projective geometry.

Note 2 Veblen also states an axiom of uniqueness of parallels (Axiom XII) which is required for affine geometry, but not for ordered geometry.

Note 3 A homogeneous coordinate system for P^3 can be specified with respect to any five points (no four of which are coplanar) such that the five points are allocated the coordinates $(1, 0, 0, 0)$, $(0, 1, 0, 0)$, $(0, 0, 1, 0)$, $(0, 0, 0, 1)$ and $(1, 1, 1, 1)$ respectively. For further details see, for example, Borsuk and Szmielew (1960).

Note 4 Projective space P^3 is categorical, as demonstrated by several authors, including Borsuk and Szmielew (1960, Chapter IX, §3).

Note 5 Busemann (1955) states and proves the proposition: "(16.11) Let $\mathcal{C}$ be a closed convex surface in A^n and z an interior point of $\mathcal{C}$. If for any two points p and q of $\mathcal{C}$ an affinity exists that leaves z fixed, maps $\mathcal{C}$ on itself and p on q, then $\mathcal{C}$ is an ellipsoid with center z.". The convex surface $\mathcal{C}$ of Busemann's proposition corresponds to the boundary $\partial \mathcal{K}$ of the bounded convex body $\mathcal{K}$ of our Theorem 82.

Notes to Chapter 10: Standard model of Minkowski space–time

Note 1 In the present treatment the value of the "speed of light" is unity as a direct consequence of the procedure for coordinatization based on equivalence classes of parallels in Section 7.3. This 1+1–dimensional coordinate system is extended, in Theorem 83 of Chapter 9, to a single 3+1–dimensional coordinate system for the entire set of events. It is only in the penultimate section of the present chapter, after a discussion of the motions of Minkowski space–time, that a larger set of coordinate frames is defined. The coordinatization procedure not only excludes the consideration of differing "units" for time and distance but also excludes the consideration of coordinate systems with "reversed" time scales. This is the reason why it is the orthochronous Poincaré group which relates the coordinate systems. In this monograph it is not the coordinate frames, or their relationships, but rather the categorical representation of the space–time which is of most interest.

Note 2 The results of Section 10.4 may be used to map arbitrary configurations on to configurations based upon the origin of space–time and the origin of position space — many of the demonstrations then become trivial.

Notes to Chapter 11: Independence models

Note 1 Veblen's axioms I–XII are categorical for affine geometry but not for Euclidean geometry, since the twelve axioms are also satisfied by a 2+1 – dimensional Minkowski space–time which can be obtained by specifying a hyperbolic polarity (instead of an elliptic polarity) in the improper plane "at infinity". Thus to specify Euclidean geometry, Veblen could have stated the existence of the elliptic polarity as an "Axiom XIII" which would then be independent of the preceding axioms. However the "Axiom XIII" would be quite complicated in its explicit statement, since it involves the existence of the improper plane at infinity and the concept of a polar correlation in this plane. Whether an "Axiom XIII" could be formulated in such a way that the entire set of thirteen axioms would be independent of each other, remains an open question.

Note 2   See Notes 3 and 4 of Chapter 2.

Note 3   By the four–dimensional extension of a result of Busemann and Kelly (1953, Theorem 13.6 and the preceding paragraph). The extension to four dimensions is obtained by applying the two dimensional result to every line through the centre a.

Note to Appendix 1: Veblen's Axioms for Ordered Geometry

Note 1   The axioms of Veblen, O. (1904), 'A system of axioms for geometry', *Transactions of the American Mathematical Society* **5** (1904), pp. 343–384 are reproduced with the kind permission of the American Mathematical Society.

Note 2   Veblen does not include any definitions named as Def. 3. or Def. 4..

Notes to Appendix 2:   Alternative axiom systems

Note 1   The statements of Axioms O5*, O6*, I6*, I7* and C* are simpler than their non–asterisked counterparts since they are stated directly in terms of the betweenness relation (rather than with a description involving the concept of a chain as defined in Section 2.2 and in Section A2.5).

The statements of Axioms O1, O2, O3, O4, I1, I2, I3, I4, I5 and S are the same as in the system of independent axioms.

Note 2   This definition is simpler and weaker than the corresponding definition for Axiom C. Thus Axiom C* is weaker than Axiom C.

Notes to Appendix 3:   Open Questions

Note 1   See Note 1 of Chapter 2.

Note 2   Many theorems corresponding to those of the present axiom system could be obtained in much the same way. However it is not known how far this deductive process could be taken. It appears that the really difficult proposition, which may not even be provable, is a "causality theorem" analogous to Theorem 26. If it turns out that a "causality theorem" can not be proved, a modified first–order axiom system would need to include some modification to at least one of the other axioms or else at least one further axiom. Then it may not be possible to establish the mutual independence of the axioms and schema of continuity axioms.

Note 3   A modified first–order axiom system would need to include some modification to at least one axiom, or one further axiom, so as to exclude models such as M_{C_3} .

Note 4 Schnabel and Pambuccian (1985) have presented a non–categorical system of eleven independent axioms for plane Euclidean geometries over Pythagorean fields.

Note 5 See Section 11.6 and the introductory comments to Section 7.5.

Index of notation

References

Aleksandrov, A. D. (sic. same author as Alexandrov, A. D.) (1969) 'Cones with a transitive group', *Dok. Akad. Nauk SSR* **189** (1969), 695-698 [*Sov. Math. Dok.* **10** (1969), 1460–1463].

Alexandrov, A. D. (sic. same author as Aleksandrov, A. D.) (1967) 'A contribution to chronogeometry', *Canadian Journal of Mathematics* **19** (1967), 1119–1128.

Borsuk, K. and Szmielew, W. (1960) *Foundations of Geometry.* Revised English translation. North Holland, Amsterdam.

Bourbaki, N. (1966) *Elements of Mathematics, Volume III Part 2, General Topology.* Addison–Wesley, Massachusetts.

Bourbaki, N. (1968) *Elements of Mathematics, Volume I, Theory of Sets.* Addison–Wesley, Massachusetts.

Browder, F.E. (1974) Mathematical Developments Arising From Hilbert Problems, In: *Proceedings of Symposia in Pure Mathematics of the American Mathematical Society* **28** , Volumes 1 and 2. American Mathematical Society, Providence.

Bunge, M. (1967) *Foundations of Physics.* Springer–Verlag, Berlin.

Busemann, H. (1955) *The geometry of geodesics.* Academic Press, New York.

Busemann, H. (1967) *Timelike spaces.* Dissertationes Mathematicae (Rozprawy Matematyczne) **53**. Warszawa, Warsaw.

Busemann, H. and Kelly, P. (1953) *Projective Geometry and Projective Metrics.* Academic Press, New York.

Carter, B. (1971), 'Causal structure in space–time', *General Relativity and Gravitation* **1** (1971), 349–391.

Castagnino, M.A. and Harari, D.D. (1982) 'Axiomatic approach to space–time geometry', *Revista de la Unin Matemtica Argentina* **30** (1982/83), 147–166.

Castagnino, M.A. and Ordóñez, A. (1989) 'Fiber bundles, sprays and the axiomatic theory of space–time', *Rendiconti di Matematica e delle sue Applicazioni. Serie VII* **9** (1989), 299–334.

Coleman, A. and Korte, H. (1984), 'Constraints on the nature of inertial motion arising from the universality of free fall and the conformal causal structure of space–time', *Journal of Mathematical Physics* **25** (1984), 3513–3526.

Coxeter, H. S. M. (1965) *Non–Euclidean Geometry.* Fifth edition. Mathematical Expositions No. 2. University of Toronto Press, Toronto.

Coxeter, H. S. M. (1961) *Introduction to Geometry.* Wiley, New York.

Ehlers, J. (1973) 'Survey of general relativity theory'. In: Israel, W. (ed) *Relativity, Astrophysics and Cosmology.* Reidel, Dordrecht, pp 1–125.

Ehlers, J. , Pirani, F.A.E. and Schild, A. (1972) 'The geometry of freee fall and light propogation'. In: O'Raifeartaigh, L. (ed) *General Relativity, Papers in Honour of J.L. Synge.* Clarendon Press, Oxford, pp63–84

Ehlers, J. and Köhler, E. (1977), 'Path structures on manifolds', *Journal of Mathematical Physics* **18** (1977) 2014–2018.

Ehlers, J. and Schild, A. (1973), 'Geometry in a manifold with projective structure', *Communications in Mathematical Physics* **32** (1973), 119–146.

Einstein, A (1905), 'Zur Elektrodynamic bewegter Körper', *Annalen der Physik* **17** (1905), 891.

Fock, V. A. (1964) *The theory of space time and gravitation.* Translated from the Russian by Kemmer, N. Second revised edition, Pergamon, Oxford.

Freudenthal, H. (1964), 'Das Helmholtz-Liesche raumproblem bei indefiniter metrik', *Mathematische Annalen* **156** (1964), 263–312.

Freudenthal, H. (1965) 'Lie groups in the foundations of geometry'. *Advances in Mathematics* **1**, 145–190.

Fulks, W. (1961) *Advanced Calculus.* Wiley, New York.

Goldblatt, R. (1987) *Orthogonality and Space–Time Geometry.* Springer, New York.

Guts, A. K. (1982), 'Axiomatic relativity theory', *Usp. Mat. Nauk* **37**:2 , 39–79. [*Russian Mathematical Surveys* **37**:2 , 41–89.].

Guts, A. K. and Levichev, A. V. (1984), 'On the foundations of relativity theory', *Dokl. Akad. Nauk* **277** (1984), 1299–1303. [*Soviet Math. Dokl.* **30** (1984), 253–257.]

Henkin, L. , Tarski, A. and Suppes, P. , (eds) (1959) *The Axiomatic Method*, North Holland, Amsterdam.

Hilbert, D. (1899) *Gründlagen der Geometrie*, first edition published in *Festschrift zur Feier der Enthullung des Gauss–Weber–Denkmals*, Teubner, Leipzig, 1899, 3–92.

234

Hilbert, D. (1900) 'Mathematical Problems', Lecture delivered before the International Congress of Mathematicians, Paris, 1900. Göttinger Nachrichten, 1900, pp253–297. Also published in Archiv der Mathematik und Physik, Third Series, **1** (1901), 44–63 and 213–237. Translated in *Bulletin of the American Mathematical Society* **8** (1902), 437–479. Also appears in English translation in Browder (1974).

Hilbert, D. (1913) *Foundations of Geometry*, translated from *Gründlagen der Geometrie*, Second Edition, Teubner, Leipzig, 1913.

Hilbert, D. (1971) *Foundations of Geometry*, Second Edition, translated by Unger, L. from the German edition, Tenth Edition, Open Court, La Salle Illinois, 1971.

Kronheimer, E. H. (1971), 'Time–ordering and topology', *General Relativity and Gravitation* **1** (1971), 261–268.

Kronheimer, E. and Penrose, R. (1967), 'On the structure of causal spaces', *Proceedings of the Cambridge Philosophical Society* **63** (1967), 481–501.

Klein, F. (1871), 'Über die sogenannte Nicht–Euclidische Geometrie', *Mathematische Annalen* **4** 1871, 573–625. This also appears in Klein (1921–1923).

Klein, F. (1921–1923) *Gesammelte mathematische Abhandlungen*, (published in three volumes), Springer, Berlin, 1921–1923, pp 573–625.

Kundt, W. and Hoffmann, B. (1962) *Determination of gravitational standard time*, in *Recent Developments in General Relativity*, PWN, Warsaw and Pergamon, Oxford, 1962.

Marzke, R.F. and Wheeler, J.A. (1964) *Gravitation as geometry I*, in Chiu, H.Y. and Hoffmann, W.F. (eds), *Gravitation and Relativity*, Benjamin, New York, 1964.

Mayr, D. (1983) *A constructive-axiomatic approach to physical space and spacetime geometries of constant curvature by the principle of reproducibility*, in Mayr, D. and Süssmann, G. (eds.), *Space Time and Mechanics*, Reidel, Dordrecht, 1983.

Milne, E. A. (1935) *Relativity, gravitation and world-structure*, Oxford University Press, Oxford, 1935.

Milne, E. A. (1948) *Kinematic Relativity*, Oxford University Press, Oxford,, 1948. [See also Milne and Whitrow (1938) and Milne (1935)].

Milne, E. A. and Whitrow, G.J. (1938) *On the meaning of uniform time, and the kinematic equivalence of the extra-galctic nebulae*, Zeitschrift für Astrophysik **15** (1938), 263–298.

Moore, E. H. (1896), 'Tactical Memoranda I–III', *American Journal of Mathematics*

18 (1896), 264–303.

Moore, R. L. (1908), 'Sets of metrical hypotheses for geometry', *Trans. Amer. Math. Soc.* **9** (1908), 487–512.

Mundy, B. (1986a), 'The physical content of Minkowski geometry', *British Journal of Philosophy of Science* **37** (1986), 25–54.

Mundy, B. (1986b), 'Optical axiomatisation of Minkowski space–time geometry', *Philosophy of Science* **53** (1986), 1–30.

Noll, W. (1964), 'Euclidean geometry and Minkowskian chronometry', *American Mathematical Monthly* **71** (1964), 129–144.

Pasch, M. (1882) *Vorlesungen über neuere geometrie*, Teubner, Leipzig, 1882.

Pasch, M. (1926) *Vorlesungen über neuere geometrie*, second edition, Springer, Berlin, 1926.

Padoa, A. (1900) *Un nouveau système irreducible de postulates pour l'algèbre*, Compte rendu du Deuxième Congrès Internationale des Mathèmaticiens tenu à Paris du 6 au 12 aôut 1900, Gauthier–Villars, 1902, 249–256.

Peano, G. (1889) *I principii di geometria logicamente esposti*, Fratelli Bocca, Torino, 1889.

Peano, G. (1894), 'Sui fondamenti della geometria', *Rivista di Matematica* **4** (1894), 51–90.

Penrose, R (1989) *The Emperor's New Mind*, Oxford University Press, Oxford, 1989.

Pimenov, R. I. (1988), 'Axiomatics of generally relativistic and Finsler space–times by means of causality', *Sibirsk. Mat. Zh.* **29** (1988), 133–143. [(English Translation) *Siberian Math. J.* **29** (1988), 267–275.].

Pirani, F.A.E. (1973), 'Building space–time from light rays and free particles', *Symposia Mathematica* **XII** (1973) 67–83, Academic Press, London, 1973. ???

Rédei, L. (1968) *Foundation of Euclidean and non–Euclidean geometry according to F. Klein*, Akadémiai Kiadó, Budapest, 1968.

Reichenbach, H. (1924) *Axiomatik der relativistischen Raumzeitlehre*, Vieweg, Braunschweig, 1924.

Robb, A. A. (1911) *Optical geometry of motion: a new view of the theory of relativity*, W. Heffer, Cambridge, 1911.

Robb, A. A. (1914) *A theory of time and space*, Cambridge U. P., Cambridge, 1914. An outline and summary appears as Robb (1921) and a revision appears as Robb (1936).

Robb, A. A. (1921) *The absolute relations of time and space*, Cambridge U. P. , Cambridge, 1921. (80p). This is an outline and summary of the axiomatic system of Robb (1914).

Robb, A. A. (1936) *Geometry of time and space*, Cambridge University Press, Cambridge, 1936. This axiomatic system is a revised edition of Robb (1914).

Russell, B. (1907) *Principles of mathematics*, Cambridge University Press, Cambridge, 1907.

Schnabel, R. and Pambuccian, V. (1985), 'Die metrisch–euklidische Geometrie als Ausgangspunkt für die geordnet–euklidische Geometrie', *Expositiones Mathematicae* **3** (1985), 285–288.

Schrödinger, E. (1956) *Expanding Universes*, Cambridge University Press, Cambridge, 1956.

Schutz, J. W. (1973) *Foundations of special relativity: kinematic axioms for Minkowski space-time*, Lecture Notes in Mathematics **361**, Springer, Berlin–Heidelberg–New York, 1973.

Schutz, J. W. (1981), 'An axiomatic system for Minkowski space–time', *Journal of Mathematical Physics* **22** (1981), 293–302.

Schutz, J. W. (1989), 'Affine structure and isotropy imply Minkowski space–time and the orthochronous Poincaré group', *Journal of Mathematical Physics* **30** (1989), 635–638.

Schutz, J. W. (1990), 'The isotropy mappings of Minkowski space–time imply the orthochronous Poincaré group', *Journal of the Australian Mathematical Society (Series B)* **31** (1990), 425–433.

Smorodinsky, J.A. (1965), 'Kinematik und Lobatschewski–geometrie', *Fortschr. Physik* **13** (1965), 157–173.

Suppes, P. and Rubin, H (1954), 'Transformations of systems of relativistic particle mechanics', *Pacific Journal of Mathematics* **4** (1954), 563–601.

Suppes, P. (1959) *Axioms for relativistic kinematics with or without parity*, in Henkin (1959), 291–307.

Suppes, P., Krantz, D.M., Luce, R.D. and Tversky, A. (1989) *Foundations of measurement, Volume II*, Academic Press, San Diego

Szczerba, L.W. (1970), 'Independence of Pasch's axiom', *Bulletin de l'Acadmie Polonaise des Sciences. Srie des Sciences Mathmatiques, Astronomiques et Physiques* **18** (1970), 491–498.

Szczerba, L.W. and Szmielew, W. (1970), 'On the Euclidean geometry without the Pasch axiom', *Bulletin de l'Acadmie Polonaise des Sciences. Srie des Sciences Mathmatiques, Astronomiques et Physiques* **18** (1970), 659–666.

Szekeres, G. (1968) 'Kinematic geometry: an axiomatic system for Minkowski space–time', *Journal of the Australian Mathematical Society* **8** (1968), 134–160.

Szekeres, P. (1994) *Discrete space–time*, in: Carey, A.L. (ed) *Confronting the infinite*, World Scientific, Singapore (1994), pp. 293–303.

Szekeres, P. (1991), 'Signal spaces – an axiomatic approach to space–time', *Bulletin of the Australian Mathematical Society* **43** (1991), 355–363.

Szmielew, W. (1980), 'Concerning the order and semi–order of *n*–dimensional Euclidean space', *Fundamenta Mathematicae* **57** (1980), 47–56.

Tarski, A. (1935), 'Einige methodologische untersuchungen über die definierbarkeit der Begriffe', *Erkenntnis* **5** (1935), 80–100. Reprinted in Tarski (1956).

Tarski, A. (1956) *Logic, semantics, mathematics, papers from 1923 to 1938*, translated by Woodger, J.H., Clarendon Press, Oxford, 1956.

Tarski, A. (1959) *What is elementary geometry?* in Henkin (1959), 16–29.

Tarski, A. (1967) *The completeness of elementary algebra and geometry*, Institute Blaise Pascal, Paris, 1967.

Torretti, R. (1978) *Philosophy of geometry from Riemann to Poincaré*, Riedel, Dordrecht, 1978.

Veblen, O. (1904), 'A system of axioms for geometry', *Transactions of the American Mathematical Society* **5** (1904), 343–384.

Veblen, O. (1911), *The foundations of geometry*, in Young, J. W. A. (1911), 3–51.

Veblen, O. and Young, J. W. (1908), 'A set of axioms for geometry', *American Journal of Mathematics* **30** (1908), 347–380.

Veblen, O. and Young, J. W. (1910) *Projective geometry*, Volume 1, Boston, 1910. Reprinted by Blaisdell, New York, 1938.

Veblen, O. and Young, J. W. (1918) *Projective geometry*, Volume 2 (by Veblen alone), Boston, 1918. Reprinted by Blaisdell, New York, 1938.

Walker, A. G. (1948), 'Foundations of relativity: Parts I and II', *Proceedings of the Royal Society of Edinburgh Section A* **62** (1948), 319–335.

Walker, A. G. (1959) *Axioms for Cosmology*, in Henkin (1959), 308–321.

Whitehead, A. N. (1907) *The axioms of descriptive geometry*, Cambridge University

Press, Cambridge, 1907.

Weyl, H. (1918) *Raum, Zeit, Materie*, Springer, Berlin, 1918. Translated as *Space, Time, Matter*,

Woodhouse, N. M. J. (1973), 'The differentiable and causal structures of space–time', *Journal of Mathematical Physics* **14** (1973), 495–501.

Yaglom, I. M. (1969) *Printsipi otnositelnosti Galileya i Neevklidova Geometriya*, Nauka, Moscow, 1969. [*A simple non–Euclidean geometry and its physical basis*, Springer, New York, 1979].

Yaglom, I. M. , Rozenfel'd, B. A. and Yosinskaya, E. U. (1964) *Russian Mathematical Surveys* **19** (1964), 49–107.

Yamasaki, H. (1983), 'Extension of trigonometric and hyperbolic functions to vectorial arguments and its application to the representation of rotations and Lorentz transformations', *Foundations of Physics* **13**, 1139–1154.

Young, J. W. A. (1911) *Monographs on topics in modern mathematics relevant to the elementary field*, Longmans Green, 1911. Reprinted by Dover, New York, 1955.

Index